RECONFIGURABLE AND ADAPTIVE COMPUTING

THEORY AND APPLICATIONS

RECONFIGURABLE AND ADAPTIVE COMPUTING

THEORY AND APPLICATIONS

EDITED BY

Nadia Nedjah

State University of Rio de Janeiro
Brazil

Chao Wang

University of Science and Technology of China
Hefei, Anhui, People's Republic of China

CRC Press
Taylor & Francis Group
Boca Raton   London   New York

CRC Press is an imprint of the
Taylor & Francis Group, an **informa** business
A CHAPMAN & HALL BOOK

CRC Press
Taylor & Francis Group
6000 Broken Sound Parkway NW, Suite 300
Boca Raton, FL 33487-2742

First issued in paperback 2018

© 2016 by Taylor & Francis Group, LLC
CRC Press is an imprint of Taylor & Francis Group, an Informa business

No claim to original U.S. Government works

ISBN 13: 978-1-138-89419-8 (pbk)
ISBN 13: 978-1-4987-3175-1 (hbk)

Contents

List of Figures

List of Tables

Preface

RECONFIGURABLE COMPUTING TECHNIQUES AND adaptive systems have attracted great interest as one of the most promising architectures for microprocessors. The origin of reconfigurable systems, also known as programmable logic devices or field programmable gate arrays (FPGAs), has evolved into today's complex system-on-chip FPGAs, dynamically reconfigurable FPGAs, and furthermore a variety of adaptive computing devices. Numerous approaches have been conducted to explore more credible metrics beyond raw performance, such as flexibility, dependability, and low power.

This book introduces and demonstrates the latest research activities on hardware architecture for reconfigurable and adaptive computing systems. The book covers three main topics: reconfigurable systems, network-on-chip, and system codesign.

Section I includes four interesting works on reconfigurable systems.

In Chapter 1, the authors present a software and hardware codesign flow for the coarse-grained systems-on-chip. The chapter enables a multi-target design space exploration (MT-DSE) algorithm with multiple objectives such as chip area utilization, energy consumption, core efficiency, interconnection structure, application workload, and speedup aware. With the help of the MT-DSE tool, the proposed design flow can provide valuable assistance to architecture designers in the development of an effective trade-off multiprocessor system. Besides, the tool is employed at a very early stage in component-based systems design and needs only a little profiling information, which can greatly reduce the development term of the design. As an illustration, a JPEG compression algorithm is chosen to demonstrate how the tool exploits a given application and guides the user to build the most desired architecture.

In Chapter 2, the authors propose the video watermarking algorithm, which is suited for the H.264 standard. The blind- and invisible-watermark-embedding scheme is implemented on a hardware platform. The algorithm is used to insert the watermark in AC coefficients of the integer DCT transform. The robustness of the video watermarking algorithm is improved by embedding the different parts of the same watermark in different frames of the same scene. An efficient architecture of the watermarking algorithm is implemented using pipeline and parallel processing for better real-time performance.

In Chapter 3, the authors describe the solution for regular expressions matching systems. With the advent of hybrid system-on-chip that integrates a CPU sub-system and a programmable logic array, the solution discussed presents a new approach to the problem. The authors introduce the original multistage algorithm for incremental data screening. The algorithm is intended for reconfigurable SoC devices, as it gains from the execution of its initial stage in the custom coprocessor. The basic idea of the algorithm comes from the concept of the Bloom filter. It requires the extraction of static substrings for strings of regular expressions. The chapter gives the architecture of pipelined custom processors and an appropriate software scheme for the accompanying CPU. Target applications for the presented solution are computer and network security systems. The idea was tested on nearly 100,000 body-based viruses from the ClamAV virus database, and the results are provided for Xilinx Zynq-7000 All Programmable system-on-chip.

In Chapter 4, the authors propose a novel FPGA-based acceleration solution with MapReduce framework on multiple hardware accelerators. The combination of hardware acceleration and MapReduce execution flow could greatly accelerate the task of aligning short length reads to a known reference genome. Also, as a practical study, this chapter builds a hardware prototype based on the real Xilinx FPGA chip. Significant metrics on speedup, sensitivity, mapping quality, error rate, and hardware cost are evaluated. Experimental results demonstrate that the proposed platform could efficiently accelerate the next generation sequencing problem with satisfying accuracy and acceptable hardware cost.

Section II comprises three interesting works on network-on-chip.

In Chapter 5, the authors propose an implementation of a multiprocessor system-on-chip platform with shared memory access. The authors investigate the performance characteristics of a parallel application running on this platform. The applications take advantage of the parallel structure of the platform, yielding very high performance.

In Chapter 6, the authors propose end-to-end quality of service metrics modeling based on a multi-application environment in network-on-chip.

The originality of their approach is based on the proposition of a quality of service intellectual property module in network-on-chip architecture to improve network performance. The results show that the quality of service modeling approach is easy to implement in a hardware–software quantifiable representation. Thus, implementing a quantifiable representation of quality of service can be used to provide a network-on-chip services arbiter.

In Chapter 7, the authors present 3D-ACR (3D ant colony routing) and apply it as a routing policy of network-on-chip having three different three-dimensional topologies: mesh, torus, and hypercube. Experimental results show that 3D-ACR performs consistently better when compared with previously proposed routing strategies.

Section III includes two interesting works on the methodology of system codesign.

In Chapter 8, the authors introduce a new software–hardware codesign flow for embedded systems, which models both processors and intellectual property cores as services. In order to guide the hardware implementations of the hot spot functions, this chapter incorporates a novel hot spot-based profiling technique to observe the hot spot functions while the application is being simulated. Furthermore, based on the hot spot of various applications, the authors present an adaptive mapping algorithm to partition the application into multiple software–hardware tasks. Experimental results demonstrate that CODEM (CODEsign Methodology) can efficiently help researchers identify hot spots. The chapter also outlines a radical new method to combine profiling techniques with state-of-the-art reconfigurable computing platforms for specific task acceleration.

In Chapter 9, the authors propose an efficient algorithm for dependent task HW–SW codesign with the greedy partitioning and insert scheduling method (GPISM) by task graph. For hardware tasks, the critical path with maximum sum of benefit-to-area ratio can be achieved and implemented in hardware while the total area occupation in this path. For software tasks, the longest communication time path can be obtained from the updated task graph and assigned for software implementation integrally. The authors conclude that the GPISM can greatly improve system performance even in the case of generation of high communication cost and efficiently facilitate researchers to partition and schedule embedded applications on multi-processor system-on-chip hardware architectures.

We are grateful to the authors and to the reviewers for their tremendous service of critically reviewing the submitted works. We thank the editorial team that helped format this work into a book. Finally, we sincerely hope that the reader will share our excitement reading this book and find it useful.

Nadia Nedjah
State University of Rio de Janeiro

Chao Wang
University of Science and Technology of China

MATLAB® is a registered trademark of The MathWorks, Inc. For product information, please contact:

The MathWorks, Inc.
3 Apple Hill Drive
Natick, MA 01760-2098 USA
Tel: 508-647-7000
Fax: 508-647-7001
E-mail: info@mathworks.com
Web: www.mathworks.com

Editors

Nadia Nedjah graduated in 1987 in systems engineering and computation and in 1990 obtained an MSc degree, again in systems engineering and computation. Both degrees were from the University of Annaba, Algeria. She obtained a PhD degree in 1997 from the University of Manchester—Institute of Science and Technology, UK. She joined the Department of Electronics Engineering and Telecommunications of the Engineering Faculty of the State University of Rio de Janeiro as an associate professor. She is currently a head of the Intelligent System research area in the electronics engineering postgraduate program of the State University of Rio de Janeiro, Brazil. Nedjah is the editor-in-chief of the *International Journal of High Performance System Architecture* and *Innovative Computing Applications*, both published by Inderscience, UK. She has published three authored books about functional and rewriting languages, hardware–software codesign for systems acceleration, and hardware for soft computing versus soft computing for hardware. She has (co)-guest edited more than 15 special issues for high-impact journals and more than 40 books on computational intelligence-related topics, such as *Evolvable Machines*, *Genetic Systems Programming*, *Evolutionary Machine Design: Methodologies and Applications,* and *Real-World Multi-Objective System Engineering.* She has (co)authored more than 90 journal articles and more than 150 conference papers. She is an associate editor of more than 10 international journals, such as Taylor & Francis Group's *International Journal of Electronics*; Elsevier's *Integration, The VLSI Journal*, and *Microprocessors and Microsystems*; and IET's *Computer & Digital Techniques*. More details can be found at http://www.eng.uerj.br/~nadia/english.html.

Chao Wang obtained a PhD degree in 2011 in computer science from the University of Science and Technology of China, Anhui, China. He was a postdoctoral researcher from 2011 to 2013 in the same university, where he is now an associate professor with the School of Computer Science. He has worked with Infineon Technologies, Munich, Germany, from 2007 to 2008. He is an associate editor of several international journals, including *Microprocessors and Microsystems*, IET's *Computer & Digital Techniques*, *International Journal of High Performance System Architecture*, and *International Journal of Business Process Integration and Management*. He has (co)-guest edited special issues for *IEEE/ACM Transactions on Computational Biology and Bioinformatics*, *Applied Soft Computing*, *International Journal of Parallel Programming*, and *Neurocomputing*. He plays a significant part in several well-established international conferences; for example, he serves as the publicity co-chair of the High Performance and Embedded Architectures and Compilers (HiPEAC) 2015 and IEEE Symposium on Parallel and Distributed Processing with Applications (ISPA) 2014; he acts as the technical program member for DATE, FPL, and FPT. He has (co)authored or presented more than 90 papers in international journals and conferences, including *IEEE Transactions on Computers*, *ACM Transactions on Architecture and Code Optimization*, *IEEE/ACM Transactions on Computational Biology and Bioinformatics*, *IEEE Transactions on Parallel and Distributed Systems*, and *IEEE Transactions on VLSI Systems*. His homepage may be accessed at http://staff.ustc.edu.cn/~cswang.

Contributors

Peng Chen
School of Computer Science
University of Science and
　Technology of China
Anhui, China

Luiza de Macedo Mourelle
Department of Systems
　Engineering and
　Computation
State University of Rio de Janeiro
Rio de Janeiro, Brazil

Luneque Del Rio Souza e Silva Jr.
Post-graduate Program of
　Systems Engineering and
　Computation
Federal University of
　Rio de Janeiro
Rio de Janeiro, Brazil

Fahui Jia
School of Software Engineering
University of Science and
　Technology of China
Anhui, China

Amit M. Joshi
Electronics & Communication
　Engineering Department
Malaviya National Institute of
　Technology (MNIT)
Jaipur, India

Chunsheng Li
School of Computer Science
University of Science and
　Technology of China
Anhui, China

Xi Li
School of Software Engineering
University of Science and
　Technology of China
Anhui, China

Xiang Ma
School of Computer Science
University of Science and
　Technology of China
Anhui, China

Vivekanand Mishra
Electronics Engineering
    Department
Sardar Vallabhbhai National
    Institute of Technology
    (SVNIT)
Surat, India

Salem Nasri
Computer and Embedded System
    Laboratory
Ecole Nationale d'Ingénieurs de
    Sfax (ENIS)
Sfax, Tunisia

and

Computer Science Department,
    College of Computer
Qassim University
Buraydah, Saudi Arabia

Nadia Nedjah
Department of Electronics
    Engineering and
    Telecommunications
State University of Rio de Janeiro
Rio de Janeiro, Brazil

R. M. Patrikar
Electronics Engineering
    Department
Visvesvaraya National Institute of
    Technology
Nagpur, India

Fábio Gonçalves Pessanha
Department of Electronics
    Engineering and
    Telecommunications
Federal University of Rio de Janeiro
Rio de Janeiro, Brazil

Paweł Russek
AGH University of Science &
    Technology
Kraków, Poland

Abdelkader Saadaoui
Computer and Embedded System
    Laboratory
Ecole Nationale d'Ingénieurs de
    Sfax (ENIS)
Sfax, Tunisia

and

Computer Science Department
ISET Radès
Radès, Tunisia

Aili Wang
School of Software Engineering
University of Science and
    Technology of China
Anhui, China

Chao Wang
School of Computer Science
University of Science and
    Technology of China
Anhui, China

Kazimierz Wiatr
AGH University of Science &
    Technology
Kraków, Poland

Qi Yu
School of Computer Science
University of Science and
    Technology of China
Anhui, China

Xuda Zhou
School of Computer Science
University of Science and
    Technology of China
Anhui, China

Xuehai Zhou
School of Software Engineering
University of Science and
    Technology of China
Anhui, China

I

Reconfigurable Systems

Effective and Efficient Design Space Exploration for Heterogeneous Microprocessor Systems-on-Chip

Chao Wang, Peng Chen, Xi Li, Xuda Zhou, Xuehai Zhou, and Nadia Nedjah

CONTENTS

1.1 INTRODUCTION

Various multiprocessor architectures are designed to enhance application parallelism and computing ability. With the continuous development of integration technology, the technology of multiprocessor system-on-chip (MPSoC) could be used to integrate substantial elements such as central processing units (CPUs), digital signal processors (DSPs), application specific instruction set processors (ASIPs), and memories on a single chip to gain more computing power [1]. It becomes possible to construct high performance and complex embedded systems on a single chip. However, this trend makes it quite hard for system designers to formulate a well-optimized architecture. In recent years, field programmable gate array (FPGA)-based coarse-grained platforms have attracted significant research interest because of their flexibility, efficiency, programmability, and many other metrics. These advantages make it easier for designers to design multiprocessor systems. Primitives in the systems do not need to be designed from the scratch. Many of the parts are reusable for different systems. What the designers need to do is to implement their particular entities and organize them as a complete computer system. This technique greatly reduces the burden for system designers. However, a large amount of work still exists for the architecture developers to cover [2]. When architecture designers begin to develop a coarse-grained architecture, they are not clear about what kind of architecture comes with the most benefit. Design space exploration tools are widely employed to tackle this problem. Factors such as the quantity of processors, interconnection structure, and memory systems are all essential issues to be concerned with while proposing a trade-off multiprocessor system. Besides, implementation via FPGA devices also imposes stringent requirements on chip area utilization, power consumption, and computing performance to the target systems. All of these elements result in a large design space and hugely increase the difficulties for system designs.

A lot of design space exploration tools have been proposed in the literature [3–6] for exploring multiprocessor architectures. However, for most of them, plenty of test cases via simulations or implementations are needed to make a choice for the trade-off architecture. Thus, a lot of time is wasted and, more seriously, the designers may sometimes fail to find the best solution. On the other side, some design space exploration techniques only focus on a particular aspect of the system [7,8], which is not enough for picking out a multiple-objective balanced architecture.

This chapter presents a software and hardware codesign flow for FPGA-based coarse-grained architectures. A novel multi-target design space exploration (MT-DSE) method is employed in the flow to help designers find the most effective trade-off architecture. Without the time-consuming simulations or implementations, this lightweight MT-DSE method only needs some early stage profile information; after that, a well-optimized scheme would be reported to the designers. System designers could benefit a lot from MT-DSE, as in recent years the requirement for time-to-market has become more and more demanding.

This chapter proposes MT-DSE tools oriented to the FPGA-based coarse-grained architectures. The design space exploration mechanism could take multiple targets into account and make a good trade-off in a short time. Except for the profiling stage, which costs several hours of time or even longer (influenced by the experience of the designers), MT-DSE can report a trade-off architecture to the designers in a matter of seconds. To illustrate the design flow and the efficiency of the MT-DSE tool clearly, we take the joint photographic experts group (JPEG) compression algorithm as the case study.

The remainder of the chapter is structured as follows: Section 1.2 describes our software and hardware codesign flow and the coarse-grained architectures in detail. Section 1.3 focuses on the mechanism of MT-DSE. Next, a consumer benchmark, the JPEG compression application, is taken as the case study, in Section 1.4. In Section 1.5, we review the related work of the design space exploration tools on multiprocessor systems. Conclusions and future work are presented in Section 1.6.

1.2 SOFTWARE AND HARDWARE CODESIGN FLOW

1.2.1 FPGA-Based Coarse-Grained Architectures

FPGA-based coarse-grained architectures stand out in the embedded systems because of their flexibility and reusability. In this chapter, we propose a design flow of this kind of architectures.

Figure 1.1 describes the basic architecture of the coarse-grained system. Several general-purpose processors are integrated for the software

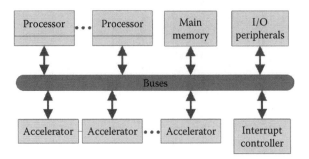

FIGURE 1.1   Coarse-grained architecture.

execution. Codes that are not suitable to execute on the hardware must be run on general-purpose processors. Memory and I/O peripherals are built in as the basic primitives in an embedded system. An interrupt controller is also available for a status reminder of special operations. Being the most beneficial components in the system, a number of hardware accelerators are used to cooperate with the processors to speed up the runtime of the applications. Usually, these accelerators are implemented to deal with some particular functions in the applications. Designers could configure different numbers and types of the elements and formulate the systems optimized for the given application. With the flexible coarse-grained architectures, the properties of each element can also be adjusted according to the requirement.

Note that the primitives are all connected with the same buses shown in Figure 1.1. In fact, the buses between different parts may be different instants and different types. For example, the I/O peripherals and the interrupt controller may share the same bus to communicate with the processors while the accelerators connect with the processors via a special high-speed link. The choice of the interconnection structure is an important design space exploration target.

1.2.2 Software and Hardware Codesign Flow

In recent years, FPGA-based coarse-grained platforms demonstrate great potential to build the trade-off multiprocessor systems because of their fine performance and scalability [9]. However, in the design of such a system, hardware platforms may change with the specifics of the applications. The platforms are not fully tested beforehand. Moreover, they may need to readjust during the development procedure. Software also should be adaptive with the hardware platforms. Therefore, a software and hardware codesign flow must ensure the correctness of the architectures. Figure 1.2 gives an overview of our software and hardware codesign flow.

When developing an embedded system for a given application, designers should always have a better understanding of the design specifications and software and hardware environments. As Figure 1.2 shows, the designers should first gather available information needed for the MT-DSE base on the application characters, the platform libraries, and the constraints. Design targets should be formulated in advance and factors that affect the efficiency of the systems should be collected based on these targets. The given application should be profiled with tools to gain some of the information and other information could be gotten from the platform libraries as well as by other ways. Then, the MT-DSE algorithm is performed to find the best trade-off architecture. For a more detailed description about the information collection and the MT-DSE, see Section 1.3.

With the help of MT-DSE, the most desired architecture is found. Thus, a development kit is used first to synthesize and implement the hardware platform. Meanwhile, software libraries should be built and the application needs to be developed. At last, software and hardware co-debug is conducted to verify the correctness of the system.

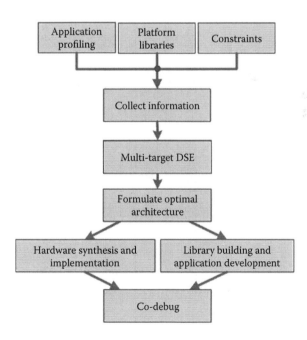

FIGURE 1.2 Codesign flow with MT-DSE.

1.2.3 Application Partition

A common but critical problem for FPGA-based embedded systems is the application partition. Many applications can be parallelized if two or more process elements are integrated into the system. Some parts in the applications can be mapped to different cores concurrently to accelerate the whole system. For FPGA-based architectures, these parts can be realized via hardware to gain a very huge speedup. Thus, the overall execution time of the application can be reduced. However, choosing parts in the applications to match the partitioned tasks requires a careful plan.

A given application can be abstracted into a serial of tasks. Here, a task refers to a function or a code block in the application. Hazards often exist in the task sequence. As an illustration, Figure 1.3 shows the tasks in an application. Tasks on the left side in Figure 1.3 are in the format of task ({input parameters}, {output parameters}). There are usually different numbers of input and output parameters for each task. On the right side of Figure 1.3, the fine arrows indicate the data dependencies within the tasks. We can learn from Figure 1.3 that task2 needs data from task1, namely, it is a read-after-write (RAW) hazard for the task2 to task1. Hence, task2 can only be executed after the task1 is completed. Those tasks that have no dependency can run in parallel, such as task3 and task4.

To enhance the performance of the whole system, every task has a possibility to be implemented via hardware. Generally speaking, those tasks with higher potential speedups are likely to be chosen as the accelerators. Each accelerator is devoted to speed up one of the tasks. There may be different kinds or quantity of accelerators integrated in the system; therefore,

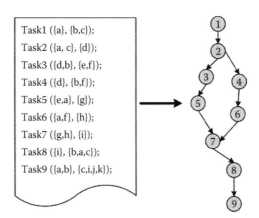

Task1 ({a}, {b,c});
Task2 ({a, c}, {d});
Task3 ({d,b}, {e,f});
Task4 ({d}, {b,f});
Task5 ({e,a}, {g});
Task6 ({a,f}, {h});
Task7 ({g,h}, {i});
Task8 ({i}, {b,a,c});
Task9 ({a,b}, {c,i,j,k});

FIGURE 1.3   Tasks in an application.

a scheduler is required to solve the task dependencies at runtime. The task partition scheme will be a challenge and also a chief design space exploration target.

1.3 MT-DSE

In this chapter, we propose a novel MT-DSE method for FPGA-based coarse-grained architectures. System designers could take benefit from it for its fast exploration and multi-target trade-off. Information collection and the multi-target exploration algorithm are two major steps in the method.

1.3.1 Information Collection

Information should be collected to perform the MT-DSE algorithm. What kind of information should be gathered is based on the design targets. Besides the two aforementioned targets (interconnection structure and task partition), there are also some other targets to be explored for the trade-off system. For example, the memory construction, the area utilization of the primitives, the power consumption, and the system performance are common issues in architecture designs. Moreover, the designers can specify their distinct targets in the constraints.

Information that has an influence on the design targets is to be gathered. This phase could be archived at the early stage of system design. To make a decision on which types of buses to use and how these buses are to be organized, information on the available buses in the platform and their characters should be present. The buses' bandwidth and their area and power consumption are all important parameters for the exploration. The features of the general-purpose processors are also a critical point. The platform-supported processor types and their maximum frequency as well as the float point computing ability are substantial references for software execution. Memory access speed and their total capacities will be useful for constituting the memory systems. Fortunately, most of the information can be found in the manufacturer datasheet; however, other information needs to be gained by profiling.

For information that cannot be gathered directly from the manufacturer datasheet, profiling tools are employed. Task partition is one of the most significant problems in embedded system designs. Which part of the application is to be realized via hardware will have a vital impact on the system performance. The partition scheme will take charge of the accelerator types and will directly affect the whole system speedup. The workload ratio of each task (the task execution time on a single processor compared to the execution time of all tasks) shows its potential to accelerate the system. High workload ratio represents more benefits in the performance enhancement. Meanwhile, once a task is selected to be implemented via hardware, it will be packaged into an IP core as the hardware accelerator. The accelerator's own speedup (the software execution time of the task compared to its hardware execution time) and chip area and power consumption are all valuable parameters to estimate the attributes of the architectures. The hardware realization of the partitioned tasks can be archived via different ways, for example, the system IP core libraries or online stores. More generally, designers can implement their own IP cores via the hardware description language or even get an instant implementation with a C to verilog translation tool.

All the above-mentioned information gathered will be used as the input of the MT-DSE algorithm. As a result, a multiple objectives balanced architecture will be explored and reported to the system designers.

1.3.2 MT-DSE Algorithm

The MT-DSE algorithm is the kernel module of the design space exploration mechanism.

Information gathered in last phase is organized with a structure shown in Figure 1.4. Types, area utilization, and power consumption are the common attributes for all primitives. The characters of capacity and the hardware/software speed are the individual issues for memory system and the accelerators. After all of these items are initialized by the information collection phase, the MT-DSE algorithm is to be performed to seek the most balanced architecture.

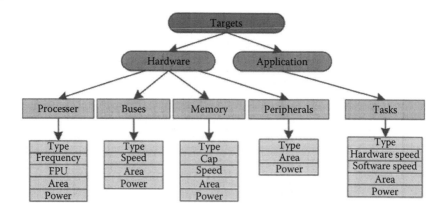

FIGURE 1.4   Information organization.

As Algorithm 1.1 shows, there are three steps in the MT-DSE totally. First, all the possible design space is searched. Every assignment in step1 represents a possible configuration of the architectures. For each primitive, the number of its instants integrated varies from zero to the max supported number. Attributes of this primitive also have different choices, which range according to the platform-supported types. Second, once all possible design space has been explored, the performance and resource consumption are calculated for every configuration. At last, the best trade-off architecture in the design space collection is picked out with the Pareto strategy. Moreover, it will be reported to the designer to finish the multi-target exploration algorithm.

ALGORITHM 1.1   MT-DSE (TARGETS, INFORMATION)

```
Begin
1.   // Step1: search all design space
2.   Collection of design space: =
3.   {
4.       For every kind of the primitives
5.       {
6.           IntegratedNumber: = [Supported Number];
7.           Attributes:= [Supported Range];
8.       }
9.   }
10.  // Step2: estimate all architectures in the
11.  Time = Information. UnPartitionedTasks. SWSpeed;
12.  Time += Information. PartitionedTasks. HWSpeed;
13.  Time += Information. Bus. TransferTime;
14.  Area = Sum[Information. Integrated. Area];
15.  Power = Sum[Information. Integrated. Power];
16.  // Step3: report the best trade-off one
17.  Trade-off Architecture:= ParetoSelect[Targets, All
     Architectures];
18.  Report Trade-off architecture;
End
```

1.4 EXPERIMENT

1.4.1 Case Study: JPEG Compression

As one of the most indispensable algorithms, the JPEG compression has great potential to be accelerated by hardware in embedded systems. It compresses a bitmap picture into a JPEG format file with little pixel loss, while the file size can be reduced dozens of times. The main body of the algorithm is a loop of the color bitmap compression with an 8 × 8 block as one unit. For each 8 × 8 block, steps illustrated in Figure 1.5 are performed.

For each recursion, an 8 × 8 Red-Green-Blue block is read from the origin bmp file first and is converted to the Y-Cr-Cb color space (CC). Then, each vector in the (Y, Cr, Cb) tuple is handed over for a two-dimensional discrete cosine transform (DCT-2D). Next, all the data in the unit are normalized (Quant) for further encoding. At last, ZigZag and Huffman (ZZ/Huffman) encoding algorithm is used to compress the block data into the final bit stream. The bit stream is written into the JPEG file at the end of the iteration.

To trickle the task partition problem, we use a tuple (Rn, An, Pn, Sn) to describe the characters of each step. *Rn* represents the workload ratio of each step (software execution time rate in the whole block procedure), *An* represents chip area cost for each IP core, *Pn* represents power consumption for IP core, and *Sn* represents speedup of the IP core itself.

Between two iterations, only the step ZZ/Huffman has the RAW dependency while the others have no hazards. To describe the task dependency more clearly, the hazard among the task sequences in the JPEG compression algorithm is illustrated in Figure 1.6. Task1, task2, task3, task4, respectively, represent the CC phase, the DCT-2D phase, the Quant phase, and the ZZ/Huffman phase.

Tasks on the left side in Figure 1.6 show the feedback loop of the JPEG compression algorithm with four RAW hazards intertasks. The right side (pointed by the dotted line) lists two unrolling loop tasks. We can see that all tasks in the first iteration have no hazards with the first three tasks in the second iteration. Only the fourth task in different iterations has the RAW hazards.

1.4.2 Hardware Platform

We have chosen the Xilinx XUPV5 board as the experimental hardware platform. Prototype architectures are implemented to demonstrate our codesign flow and test its performance.

Figure 1.7 presents one of the prototype architectures with one processor and one accelerator integrated. The MicroBlaze CPU is chosen as the

FIGURE 1.5   JPEG 8 × 8 block process steps.

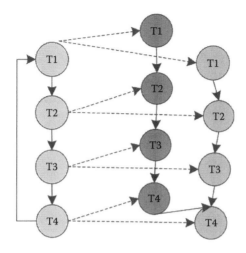

FIGURE 1.6   JPEG interloop tasks dependencies.

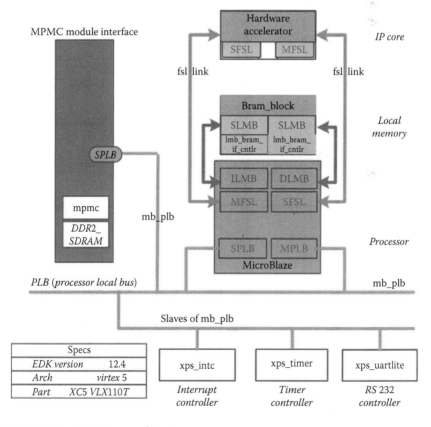

FIGURE 1.7   Prototype architecture.

general-purpose processor. An optional float process unit has been integrated in the MicroBlaze core. Peripherals such as the interrupt controller, the timer controller, and the RS232 controller are connected to the processor with the same processor local bus (PLB). An extended DDR2 memory is also connected to the processor with the PLB. Specially, a small on-chip local memory is implemented to provide faster access speed for the processor. Besides, a pair of fast simple link (FSL) [10] is used to connect the IP core and the MicroBlaze for high-speed communication. All these primitives together compose a full computer system. The numbers of the processors and accelerators could be adjusted according to the requirement of the given applications.

To enhance the programmability of the system, all the accelerators (the hardware realization of the partitioned tasks) are packaged with a uniform framework. Thus, the accelerators can be easily integrated into the system. Software handles for these accelerators can be built in with a uniform interface as well. The IP core framework is shown in Figure 1.8.

The framework communicates with other parts with a first in first out (FIFO) interface, which is adaptive for different data amount and transfer rate [11]. Numerals on the lines indicate the data width of the signals. At the beginning, input data are gotten from the FIFO interface and cached in the input buffer. Next, the buffered input data are sent to IP core to process. Results are sent to the output buffer and stored until all data are available. At last, the result is communicated to other primitives through the FIFO interface. All these operations are under the control of the state machine logic.

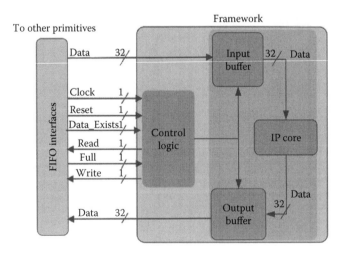

FIGURE 1.8   Accelerator package framework.

1.4.3 Information Collection Phase

Profiling is made by tools to evaluate the characters of each task in the JPEG encoding. Application workload of different steps in one 8 × 8 block is evaluated on a single-processor platform. The average ratio for CC, DCT-2D, Quant, and ZZ/Huffman in each block is 49.73%, 40.8%, 7.51%, and 1.94%, respectively, which is listed in Table 1.1. It is different from the version in Wu et al. [5]. This is reasonable as applications coded by different programmers always differ with each other. Different hardware platforms will also have a large effect on the execution speed of the application tasks. It should not matter for a fine-designed space exploration algorithm to deal with different implementation entities.

Table 1.1 lists the profile information of the JPEG compression algorithm. In the experiment, only steps of CC, DCT-2D, and Quant are packaged as the accelerators. The step of ZZ/Huffman, in contrast, has a rather too low workload ratio (1.94%) that it has little potential to be realized via hardware. Hardware speed and software speed in the table describe the average cycles cost by each step in every block. We can calculate the partial speedup of the accelerators with the two rows' data. The whole speedup can also be calculated if associating with the workload ratio information. LUTs, Flip Flops, and the BRAMs are all resources integrated into the FPGA chip. The number occupied by systems can be evaluated with Xilinx EDK tools. The area chip consumption can be calculated with this information [12]. Besides, the consumed energy can also be estimated. Data in Table 1.1 are the statistics result at a 100 MHz clock rate.

Table 1.2 gives information about the buses and memories supplied in the hardware platform. The PLB, the local memory bus, and the FSL are

TABLE 1.1   Application Profiling Result

Items	CC	DCT-2D	Quant	ZZ/Huffman
HWSpeed	257	1263	70	–
SWSpeed	588220	47767	8800	2277
Ratio	49.73	40.8	7.51	1.94
LUTs	36	44	44	–
Flip Flops	33	7	7	–
BRAMs	0	0	0	–
Area	0.133	0.204	0.117	–
Power	2.53	6.76	2.17	–

Note: The unit for the HW speed and SW speed is cycles; for ratio, percentage; for LUT, FF, and BRAM, numbers; for area, mm^2; and for power, mW.

TABLE 1.2   Information about the Bus and Memory

	Bus			Memories	
	PLB	**LMB**	**FSL**	**DDR2**	**LocalRAM**
LUTS	559	353	3	3488	0
Flip Flops	0	0	36	2062	0
BRAMs	0	0	0	5	Volatile
Capacity	–	–	–	56MB	512K
Power	0.12	0.01	0.04	100.85	Volatile

Note: The unit for the LUT, FF, and BRAM is number; for power, mW.

common buses used in the Xilinx systems. On the other side, the system memory structure mainly contains the on-chip local RAM and the extended DDR2 memory. The data in Table 1.2 show the resources occupied by each primitive of the instant. They are profiled or looked up from the manufacturer datasheet. The capacity of the memories represents the total amount that the platform supports. Recourses cost by the local RAM are volatile with the final integrated size.

1.4.4 Design Space Exploration Phase

MT-DSE is decided according to the data collected during the information collection phase. Design targets of the interconnection structure and the task partition have already been deterministic based on the experiment platform and the profiling results. The system speedup, the chip area utilization, the energy consumption, and the core efficiency are chosen as the design targets to make a typical balanced architecture. Here, core efficiency is defined [5] in Formula 1.1 as follows:

$$\text{Core efficiency} = \frac{\text{Speedup}}{\text{Core number}} \tag{1.1}$$

The targets and the information gathered are taken as the input of the MT-DSE algorithm. Then the trade-off architecture will be reported to the designer. We list all the explored architectures here to show the details in the design space exploration procedure.

Performances of explored architectures are shown in Figure 1.9, in which the speedup, full system power consumption, chip area cost, and the core efficiency are reported. Title of each column in Figure 1.9 indicates the configuration of the architecture. For example, 4MB indicates a homogeneous system with 4 MicroBlaze CPUs, and 1MB + 1CC + 1Quant indicates a hybrid system with one MicroBlaze CPU, a color covert accelerator, and a Quant accelerator.

	1MB	2MB	3MB	4MB	1MB+1CC	1MB+1DCT-2D	1MB+1Quant	1MB+1CC+1DCT-2D	1MB+1CC+1Quant	1MB+1Quant+1DCT-2D	1MB+1CC+1DCT-2D+1Quant
Speedup	1	1.96	2.89	3.78	1.98	1.66	1.08	10.28	2.33	1.89	11
Efficiency	1	0.98	0.96	0.94	0.99	0.83	0.54	3.43	0.78	0.63	7.58
Area^a	4.23	5.96	7.69	9.42	4.37	4.44	4.35	4.57	4.48	4.56	4.69
Power^b	262.23	292.77	323.3	353.84	264.76	268.99	264.4	271.52	266.93	271.8	273.69

a: unit in mm^2, b: unit in mW.

FIGURE 1.9 Performance of estimated architectures.

By observing only the first four architectures in Figure 1.9, we know that homogeneous architectures can obtain an increasing speedup (1.96 for 2MB, 2.89 for 3MB, and 3.78 for 4MB) with the increase of CPU number; however, the chip area cost (from 4.23 to 9.42 mm²) also increases rapidly. On the other side, it can be noticed that the core efficiency changes from 1 to 0.94 and power consumption changes from 262.23 mW to 272.73 mW. On the other side, the last seven architectures are situations of the hybrid architectures. In marked contrast with the homogeneous architectures, the result for hybrid architectures demonstrates that they could archive significant speedup up to 30.3 with little power consumption (varies between 264.4 mW and 271.52 mW) and chip area cost (varies between 264.76 mm² and 271.52 mm²). As illustrated in Figure 1.9, the peak speedup of 30.3 is archived with the configuration of 1MB + 1CC + 1DCT-2D + 1Quant. The lowest speedup of 1.08 occurs at the configuration of 1MB + 1Quant (not including the 1MB base architecture). Core efficiency also comes up to the peak of 7.58 for 1MB + 1CC + 1DCT-2D + 1Quant. It is very clear that the hybrid architectures have prominent dominance than the homogeneous architectures in chip area cost and power consumption. As to the speedup, some of the hybrid architectures also have absolute advantages compared to the homogeneous architectures. Obviously, the configuration of 1MB + 1CC + 1DCT-2D + 1Quant is the most satisfying one because of its high acceleration performance and relatively low chip area and power consumption.

1.4.5 Result Analysis

To demonstrate the correctness of our design flow and the MT-DSE method, a series of architectures are built and tested in the real platforms. We conduct experiments with different configurations on the Xilinx XUPV5 board. The actual performances of the architectures are illustrated in Figure 1.10, while Figure 1.11 depicts the difference between actual architectures and the estimated ones (|estimated − actual|/actual * 100%). As in Figure 1.9, the column titles in Figures 1.10 and 1.11 also indicate the configurations of different architectures.

From Figure 1.10, we can learn that the actual architectures come up to a speedup of 26.01 for hybrid systems and a speedup of 3.83 for homogeneous systems. Based on Figure 1.11, the difference in chip area cost is less than 5.6% and that for power consumption is less than 6.8% for the homogeneous architectures. Besides, speedup and core efficiency difference all

	1MB	2MB	3MB	4MB	1MB+1CC	1MB+1DCT-2D	1MB+1Quant	1MB+1CC+1DCT-2D	1MB+1CC+1Quant	1MB+1DCT-2D+1Quant	1MB+1CC+1DCT-2D+1Quant
Speedup	1	2.02	2.96	3.83	1.97	1.65	1.08	8.99	2.30	1.88	10
Efficiency	1	1.01	0.99	0.96	0.99	0.83	0.54	3	0.77	0.63	6.5
$Area^a$	4.23	5.63	7.29	8.92	4.44	4.47	4.37	4.67	4.57	4.6	4.8
$Power^b$	262.23	272.92	318.81	343.08	266.84	269.13	264.58	273.74	269.19	271.48	276.09

a: unit in mm^2, b: unit in mW.

FIGURE 1.10 Performance of the actual architectures.

	1MB	2MB	3MB	4MB	1MB+1CC	1MB+1DCT-2D	1MB+1Quant	1MB+1CC+1DCT-2D	1MB+1CC+1Quant	1MB+1Quant+1DCT-2D	1MB+1CC+1DCT-2D+1Quant
Speedup	0	3%	3%	1%	0	0	0	13%	1%	0	14%
Efficiency	0	3%	3%	1%	0	0	0	13%	2%	0	14%
Area[a]	0	6%	5%	5%	2%	1%	0	2%	2%	1%	2%
Power[b]	0	7%	1%	3%	1%	0	0	1%	1%	0	1%

FIGURE 1.11 Difference between the actual and the estimated architectures.

are within 2.9%. For hybrid architectures, all items have little difference except for the speedup and efficiency of 1MB + 1CC + 1DCT-2D and 1MB + 1CC + 1DCT-2D + 1Quant. Both of them have a difference up to 14.2%. The Quant and ZZ/Huffman steps in the JPEG compression 8 × 8 block only take a small ratio together. Therefore the whole block execution time is very short for these two configurations; so, any unmeasured delays will have a scaled influence on the system speedup.

In conclusion, the performances of the estimation architectures are quite exact compared to the real situations.

1.5 RELATED WORK

Most of the state-of-the-art design space exploration tools rely on different kinds of simulations or implementations. Researchers from Xilinx Inc. and University of Southern California [13] proposed a design space exploration technique for configurable multiprocessor platforms. Hardware and software simulators were tightly integrated to concurrently simulate the arithmetic behavior of the multiprocessor platform. Genbrugge and Eeckhout [14] describe statistical simulation as a fast simulation technique for chip multiprocessor design space exploration. By modeling the memory address stream behavior in a more micro architecture independent way and by modeling a program's time-varying execution behavior, the statistical simulation has been highly enhanced. Beltrame et al. [15] present an efficient technique to perform design space exploration of a multiprocessor platform that minimizes the number of simulations needed to identify a Pareto curve with metrics like energy and delay. The domain knowledge derived from the platform architecture is used to set up the exploration as a discrete-space Markov decision process. The famous MULTICUBE project [3,8,16,17] also takes benefit from a serial of simulators, including MULTICUBE-SCOPE, the IMECTLM platform, the STMicroelectronics SP2 simulator, and the Institute of Computing Technology's many core simulator. Aiming at leverage existing modeling, analysis, and DSE tools, Yang et al. [2], Trcla et al. [18], and Basten et al. [19] have introduced the Octopus Design-Space Exploration toolset to support model-driven design space exploration for embedded systems. It integrates CPN tools for stochastic simulation of timed systems and Uppaal for model checking and schedule optimization.

These are all perfect works on exploration of the design space for embedded systems. However, simulations of a full system consume quite a lot of time, but much less for the implementations. Time seems more and more urgent for embedded system developments under the pressure of more fierce competition and need for shorter time-to-market. To this end, a time-saving and efficient design space exploration method is strongly needed for the modern MPSoC architecture.

If the exploration cycles for the time being are excluded from consideration, some of the design space exploration flows mainly focus on one certain objective [4,7,20,21] while some others focus on more objectives [2,3,8,15–19,22,23]. Bossuet, Gogniat, and Philippe [21] focus on communication costs most when exploring the design space for reconfigurable architectures. They also came up with a functional model to describe the architectures that the designer wanted to compare. As memory system is often a critical component for multiprocessor systems, Jason et al. [7] describe a fast, accurate technique to estimate an application's average memory latency on a set of memory hierarchies. Hoang et al. [20] introduce FlexTools, a tool framework built around the FlexCore architecture, to evaluate the system performance with special attention on energy efficiency for different applications. Sotiropoulou and Nikolaidis [4] explore algorithm partitioning and system architectures for exploitation of both data and task-level parallelism and include in their study the parameter of different types of memory architectures offered on an FPGA. Ludovici et al. [23] describe an exploration method on a GALS system where the network-on-chip (NoC) and its end nodes have independent clocks (unrelated in frequency and phase) and are synchronized via dual-clock FIFOs at network interfaces. Factors such as cost efficiency and flexibility are regarded as challenge problems. Weis et al. [22] describe the design space exploration of 3D-stacked DRAMs with respect to performance, energy, and area efficiency for densities from 256Mbit to 4Gbit per 3D-DRAM channel. A previous study of this chapter was presented in Chen et al. [24], which discusses a design space exploration method for coarse-grained system-on-chip architectures. The idea is inherited in this chapter and the detailed explanation as well as the experiment is extended.

Our proposed method has both of the appealing features of time-saving and multiple targets balanced. It could turn out a desired system very fast, while a collection of targets would be analyzed for the best trade-off architecture at the same time.

1.6 CONCLUSIONS

Design space exploration plays an important role in the embedded system designs. FPGA-based platforms always have a strict limitation on available resource (chip area, energy supplied, and computational performance). To overcome the challenge of design space exploration, in this chapter we propose a hardware and software codesign flow with a MT-DSE mechanism. Multiple targets are concerned to make a balance among the chip area utilization, power consumption, interconnection structure, accelerators speedup, memory systems, and the application workload ratio. Compared to the design space exploration method in the literature, our exploration algorithm has the metric of fast speed and ease of use. Getting rid of many simulations or implementations, the MT-DSE method only needs some early stage profiling information. After that, it can report the multi-target trade-off architecture to the designers in several seconds.

JPEG compression algorithm is chosen as the case study, and a series of architectures are built with the real FPGA platform to demonstrate the correctness and efficiency of the method. As a lightweight design space exploration method, the multi-target flow described in this chapter is outstanding for its time-saving and multi-target trade-off, which we believe will bring a lot of benefits to the system designers for the fast development of products. Dynamic reconfiguration is an emerging filed in FPGA-based system design. Studies on design space exploration for reconfigurable architectures have already been presented in literature [25,26]. The extension of our multi-target design exploration method on coarse-grained reconfiguration systems stays as the future work.

FUNDING

This work was supported by the National Science Foundation of China [61379040], [61272131], [61202053], Jiangsu Provincial Natural Science Foundation [SBK201240198], Open Project of State Key Laboratory of Computer Architecture, Institute of Computing Technology, Chinese Academy of Sciences [CARCH201407], and the Strategic Priority Research Program of CAS [XDA06010403].

REFERENCES

1. C. Wang, X. Li, J. Zhang, P. Chen, Y. Chen, X. Zhou, and R. C. C. Cheung. Architecture support for task out-of-order execution in MPSoCs. *IEEE Transactions on Computers*. Vol. 99, no. PrePrints, p. 1, 2014.
2. Y. Yang, M. Geilen, T. Basten, S. Stuijk, H. Corporaal. Automated bottleneck-driven design-space exploration of media processing systems. In *Design, Automation and Test in Europe Conference and Exhibition*, Dresden, Germany, 1041–1046, 2010.
3. G. Mariani, A. Brankovic et al. A correlation-based design space exploration methodology for multi-processor systems-on-chip. In *Design Automation Conference, 47th ACM/IEEE*, Anaheim, CA, 2010.

4. C. L. Sotiropoulou and S. Nikolaidis. Design space exploration for FPGA-based multiprocessing systems. In *Proceedings of the 17th IEEE International Conference on Electronics, Circuits, and Systems,* Athens, Greece, 2010.

5. J. Wu, J. Williams et al. Design exploration for FPGA-based multiprocessor architecture: JPEG encoding case study. In *Proceedings of the 17th IEEE Symposium on Field Programmable Custom Computing Machines,* Napa, CA, 2009.

6. A. A. A. Rahman, R. Thavot, S. C. Brunet, E. Bezati, and M. Mattavelli, Design space exploration strategies for FPGA implementation of signal processing systems using CAL dataflow program. In *Proceedings of the Conference on Design and Architectures for Signal and Image Processing,* Karlsruhe, Germany, pp. 1,8, October 23–25, 2012.

7. D. H. Jason, W. D. Jack, and B. W. David, Fast, accurate design space exploration of embedded systems memory configurations. In *Proceedings of the ACM Symposium on Applied Computing,* ACM: Seoul, Korea. p. 699–706, 2007.

8. A. Gellert, G. Palermo et al. Energy-performance design space exploration in SMT architectures exploiting selective load value predictions. In *Design, Automation and Test in Europe Conference and Exhibition,* Dresden, Germany, 2010.

9. C. Wang, X. Li, J. Zhang, X. Zhou, and X. Nie. MP-Tomasulo: A dependency-aware automatic parallel execution engine for sequential programs. *ACM Transactions on Architecture and Code Optimization.* Vol. 10, no. 2, pp. 1–24, May 2013.

10. J. A. Williams, I. Syed et al. A reconfigurable cluster-on-chip architecture with MPI communication layer. In *Proceedings of the 14th Annual IEEE Symposium on Field-Programmable Custom Computing Machines,* Napa, CA, 2006.

11. C. Wang, X. Li, J. Zhang, X. Zhou, and A. Wang. A star network approach in heterogeneous multiprocessors system on chip. *The Journal of Supercomputing.* Vol. 62, no. 3, pp. 1404–1424, 2012.

12. L. Kuon and J. Rose. Area and delay trade-offs in the circuit and architecture design of FPGAs. In *Proceedings of the 16th International ACM/SIGDA Symposium on Field Programmable Gate Arrays,* ACM: Monterey, CA, pp. 149–158, 2008.

13. J. Z. Ou and V. K. Prasanna. Design space exploration using arithmetic-level hardware-software cosimulation for configurable multiprocessor platforms. *ACM Transactions on Embedded Computing Systems.* Vol. 5, no. 2, pp. 355–382, 2006.

14. D. Genbrugge and L. Eeckhout. Chip multiprocessor design space exploration through statistical simulation. *IEEE Transactions on Computers.* Vol. 58, no. 12, pp. 1668–1681.

15. G. Beltrame, L. Fossati, and D. Sciuto. Decision-theoretic design space exploration of multiprocessor platforms. *IEEE Transactions on Computer-Aided Design of Integrated Circuits and Systems.* Vol. 29, no. 7, pp. 1083–1095, 2010.

16. S. Cristina, F. William et al. MULTICUBE: Multi-objective design space exploration of multi-core architectures, in *Proceedings of the 2010 IEEE Annual Symposium on VLSI2010*, IEEE Computer Society, Lixouri Kefalonia, Greece, pp. 488–493, 2010.

17. G. Mariani, P. Avasare et al. An industrial design space exploration framework for supporting run-time resource management on multi-core systems. In *Design, Automation and Test in Europe Conference and Exhibition*, Dresden, Germany, 2010.

18. N. Trcka, M. Hendriks et al. Integrated model-driven design-space exploration for embedded systems. In *Proceedings of the 2011 International Conference on Embedded Computer Systems, Samos, Greece.*

19. T. Basten, E. V. Benthum et   al., Model-driven design-space exploration for embedded systems: The Octopus toolset. In *Proceedings of the 4th International Conference on Leveraging Applications of Formal Methods, Verification, and Validation—Volume Part I*, Springer-Verlag: Heraklion, Greece, pp. 90–105, 2010.

20. T. T. Hoang, U. Jalmbrant et al. Design space exploration for an embedded processor with flexible datapath interconnect. In *Proceedings of the 21st IEEE International Conference on Application-Specific Systems Architectures and Processors*, Rennes, France, 2010.

21. G. Gogniat, L. Bossuet, and J. L. Philippe. Communication costs driven design space exploration for reconfigurable architectures. In *Proceedings of the International Conference on Field Programmable Logic and Applications*, Lisbon, Portugal, 2003.

22. C. Weis, N. Wehn et al. Design space exploration for 3D-stacked DRAMs. In *Design, Automation and Test in Europe Conference and Exhibition*, Grenoble, France, 2011.

23. D. Ludovici, A. Strano et al. Design space exploration of a mesochronous link for cost-effective and flexible GALS NOCs. In *Design, Automation and Test in Europe Conference and Exhibition*, Dresden, Germany, 2010.

24. P. Chen, C. Wang, X. Li, and X. Zhou. Multi-objective aware design flow for coarse-grained systems on chip. *Proceedings of the 20th IEEE International Conference on Embedded and Real-Time Computing Systems and Applications*, Chongqing, China, pp. 1–8.

25. Y. Kim, R. N. Mahapatra, and K. Choi, Design space exploration for efficient resource utilization in coarse-grained reconfigurable architecture. *IEEE Transactions on Very Large Scale Integration (VLSI) Systems*. Vol. 18, no. 10, pp. 1471–1482, 2010.

26. R. Kumar and A. Gordon-Ross. Formulation-level design space exploration for partially reconfigurable FPGAs. In *Proceedings of the International Conference on Field-Programmable Technology*, New Delhi, India, 2011.

CHAPTER 2

Integer DCT-Based Real-Time Video Watermarking for H.264 Encoder

Amit M. Joshi, Vivekanand Mishra, and R. M. Patrikar

CONTENTS

2.1 INTRODUCTION

Over the years, there has been a tremendous growth in the amount of multimedia content like video, audio, and image being exchanged over computer network. With increasing processing power and bandwidth available to the users, video has emerged as the most popularly shared multimedia object. This in turn has resulted in major security concerns, with the last two decades witnessing significant research focuses (Voloshynovskiy et al. 2001; Piva et al. 2002). The technology transition from analog to digital has broadened up the gap of security and transmission for digital content. This is further worsened by the wide availability of various editing tools available, which may be used to modify the original content with ease (Lee 2003). Various countermeasures have evolved to prevent and detect such unauthorized modification. Watermarking is one such technique that can be useful to resolve ownership problem of original content.

Digital video watermarking is the art of hiding ownership data in video such that an originator can prove his/her authenticity whenever required. H.264, also known as MPEG-4 Part 10 advanced video codec, is one of the latest standards, developed jointly by ITU-T video coding expert group and ISO/IEC & Motion Picture Expert Group (MPEG) (Richardson 2004). It has a high data rate and also provides better picture quality. The compression efficiency is two/four times than MPEG-2/MPEG-4 coders, respectively (Wiegand et al. 2003). H.264 is a useful standard for cable TV and HDTV for Digital Video Broadcasting Handheld (DVB-H) systems with small screens (Dengpan et al. 2012). Real-time video watermarking is desirable for applications such as copyright protection, video authentication, video fingerprinting, copy control, and broadcast monitoring. Most of the work done in video watermarking till now was focused on uncompressed domain watermarking whereas existing electronic devices like digital camera and mobile phones provide output in one of the video compression standards. In uncompressed domain watermarking, the video is decompressed in order to embed the watermark, and then watermark embedding algorithm is run on it, which is used to embed the watermark in original video that is again recompressed to produce a secure compressed stream. For real-time applications, software watermarking is not an ideal choice as it introduces substantial delay in generating the watermark-embedded video. All software-based algorithms can be targeted on digital signal processor (DSP) implementation, but this approach is not suitable in case of real-time applications. Real-time video watermark embedding can be achieved where a custom integrated circuit (IC) resides in electronic appliances, which embeds the watermark at the time when video is captured. The embedding unit should be a part of the encoder for real-time watermark embedding. The design of a watermarking chip involves trade-offs among various performance parameters such as power, area, and real-time performance (Mathai et al. 2003b). In the present work, an algorithm with pipeline and parallel architecture is proposed, which has better speed suited for real-time requirements (Kougianos et al. 2009).

The organization of the chapter is as follows: Section 2.2 provides a brief review of existing work on real-time video watermarking. Section 2.3 describes the proposed video watermarking algorithm, which is followed by architectural design and implementation of the proposed algorithm in Section 2.4. Section 2.5 presents an analysis of the result of proposed algorithm along with comparisons with other similar works. Finally, concluding remarks appear in Section 2.6.

2.2 RELATED WORK

Digital watermarking came into existence around early 1990s for various applications. In the beginning, the watermarking algorithms were focused for images and were developed and implemented on software platforms. The watermarking algorithm requires some attributes like robustness, invisibility, and blind detection. The computational complexity is a vital for very large scale integration (VLSI) implementation (Jayanthi et al. 2012).

The real-time implementation of digital watermark embedding and detection for broadcast monitoring is demonstrated in Strycker et al. (2000). The method was implemented on Trimedia processor at 4 billion operations per second (BOPS). In watermark embedding process, a pseudo noise pattern is added to incoming video stream. The watermark gets embedded in uncompressed video and spatial domain correlation is applied at the detection side. The algorithm suffers from heavy computational load and hence is not suitable for real-time applications.

Millennium watermarking scheme for DVD copyright protection is implemented in Maes et al. (2000). The algorithm uses encryption with watermark embedding to have an extra level of security. The memory requirement for watermark embedding is huge, and this prevents its integration as a custom IC for consumer devices.

VLSI implementation of video watermark embedding unit and the detector for Just Another Watermarking System method is presented by Mathai et al. (2003a). The chapter includes floating data path architecture, filtering process, and fast fourier transform (FFT) for transformation at detection side. The method works on raw video and thus requires full encoding/decoding to embed the watermark for compression standard.

The spread spectrum-based watermarking, used to embed the watermark in both spatial and transform domain efficiently, is developed by Tsai and Wu (2003). In this work, a VLSI architecture is designed, which can be adapted for MPEG encoder. However, this encoder requires the use of floating point arithmetic, which is not suited for real-time applications.

The traceable watermarking hardware, where the watermark is inserted in the lower band of wavelet for each frame of video, is described in Vural et al. (2005). Frame selection for watermark embedding is done according to random number generation. At the encoder side, watermark is used for secure transmission along the channel. At the decoder side, watermark embedding is used to find where and when the content is manipulated. The extraction process of watermarking algorithm is blind, that is, no requirement of original content. The algorithm uses wavelets for frequency domain transformation while most video standards have DCT for the same. This incompatibility prevents the integration of watermark embedding architecture with video standard.

Perceptual watermarking concept of real-time application for video broadcasting is introduced by Mohanty et al. (2009). VLSI architecture for an MPEG-4 compression standard was designed and prototyped on field-programmable gate array (FPGA) for hardware implementation. However, the algorithm embeds a visible watermark, which restricts its utility for practical applications.

The invisible and a semi-fragile watermark embedding of compressed stream for video authentication is suggested by Roy et al. (2011). Hardware implementation of the proposed video watermarking was adapted for DCT-based MJPEG compression. The architecture was implemented on an Altera Cyclone FPGA device. Since the algorithm was based on DCT, which has floating point computation, it was not suited for H.264-based integer arithmetic.

The proposed algorithm incorporates Integer DCT, which has integer arithmetic to reduce computational complexity. In this chapter, a custom IC implementation of the algorithm is explained so that watermark embedding unit can be made an integral part of H.264 encoder. The concept of scene change detection is introduced where different bit planes of the watermark are embedded in different frames of a particular scene to improve robustness against temporal attacks.

2.3 PROPOSED VIDEO WATERMARKING USING INTEGER DCT

The proposed algorithm involves Integer DCT for spatial to frequency domain transformation. Integer DCT is the crucial module of forward path for latest H.264 video compression standard having simple arithmetic computations that are helpful in developing real-time watermark embedding process (Richardson 2010). In this chapter, a new 8×8 Integer DCT is used for better performance (Malvar et al. 2003). For a 2-D Forward DCT, the mathematical model is given by Equation 2.1

$$Y(u,v) = C(u)C(v)\left[\sum_{m=0}^{(N-1)}\sum_{n=0}^{(N-1)} X(m,n)\cos\frac{(2m+1)u\pi}{2N}\cos\frac{(2n+1)v\pi}{2N}\right] \quad (2.1)$$

where:

$$c(u) = c(v) = \frac{1}{\sqrt{N}}, \quad \text{for } u,v = 0$$

$$c(u) = c(v) = \sqrt{\frac{2}{N}}, \text{ for } u, v = 1 \text{ through } N-1; \text{ where } N = 4, 8, \text{ and } 16$$

2-D Integer DCT is calculated by separable property in two steps by successive 1-D operations carried out on each column and each row, respectively. The resultant coefficient forms an $N \times N$ matrix as shown in Equation 2.2:

$$A(i, j) = C(i)\cos\left(\frac{(2j+1)i\pi}{2N}\right) \quad (2.2)$$

The resultant matrix is defined as C_f, which is derived from $A(i, j)$ for respective value of i and j for $N = 8$. The resultant matrix is

$$C_f = \begin{bmatrix} 8 & 8 & 8 & 8 & 8 & 8 & 8 & 8 \\ 12 & 10 & 6 & 3 & -3 & -6 & -10 & -12 \\ 8 & 4 & -4 & -8 & -8 & -4 & 4 & 8 \\ 10 & -3 & -12 & -6 & 6 & 12 & 3 & -10 \\ 8 & -8 & -8 & 8 & 8 & -8 & -8 & 8 \\ 6 & -12 & 3 & 10 & -10 & -3 & 12 & -6 \\ 4 & -8 & 8 & -4 & -4 & 8 & -8 & 4 \\ 3 & -6 & 10 & -12 & 12 & -10 & 6 & -3 \end{bmatrix} \cdot \frac{1}{8} \quad (2.3)$$

As can be seen from Equation 2.3, C_f is an 8×8 matrix of 1-D Integer DCT coefficient. The matrix contains coefficients having integer values. Hardware implementation of DCT for integer values outperforms the same operation carried on floating point values. The algorithm for H.264 encoder is designed to take due advantage of 2-D DCT simplicity and compatibility with integer operation.

2.3.1 Watermark Embedding Algorithm

Input: Uncompressed video, watermark image

Output: Watermarked compressed video

Step 1: The scene change detection of video is calculated with the help of histogram difference as follows:

$$H = \sum_{i,j=0} \left[P_1(i) - P_2(j) \right]^2 \tag{2.4}$$

where:
   P_1 and P_2 are two consecutive frames

Step 2: The watermark is an 8-bit image that passes through the bit plane slice module to yield eight different planes from LSB to MSB. Each of these planes is treated as an individual watermark.

Step 3: Bit planes of the watermark are embedded into the different frames of a scene of video as shown in Figure 2.1. This approach has the advantage that if distortion occurs in any one of the frames, then the other parts of the watermark can still be retrieved from remaining frames.

Step 4: Each frame of a particular scene from the video is divided into 8×8 blocks, and Integer DCT is calculated to get one DC and different AC values.

Step 5: AC values are estimated with the aid of surrounding DC value of the blocks as shown in Figure 2.2. Prediction of AC values is carried out with the help of Equations 2.5 through 2.9.

$$AC'(0,1) = c_1 * (DC_4 - DC_2) \tag{2.5}$$

$$AC'(1,0) = c_1 * (DC_2 - DC_8) \qquad (2.6)$$

$$AC'(0,2) = c_2 * (DC_4 - DC_6 - 2 * DC_5) \qquad (2.7)$$

$$AC'(2,0) = c_2 * (DC_2 + DC_8 - 2 * DC_5) \qquad (2.8)$$

$$AC'(1,1) = c_3 * (DC_1 + DC_9 - DC_3 - DC_7) \qquad (2.9)$$

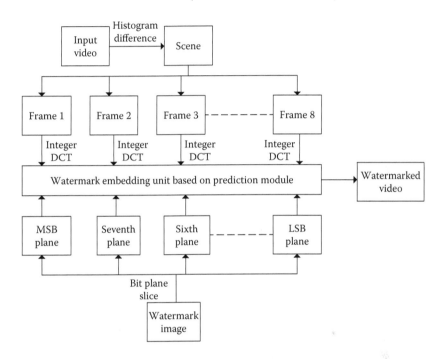

FIGURE 2.1 Proposed video watermarking method.

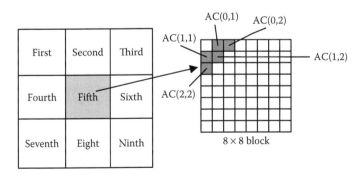

FIGURE 2.2 Prediction modules with adjacent blocks.

where:

$$c_1 = \frac{9}{64}, c_2 = \frac{1}{32}, c_3 = \frac{3}{128}$$

Here, c_1, c_2, and c_3 are constants used for AC prediction.

The constants used for AC estimation have denominator of 2^n, where n is a natural number. These values are calculated with a simple shift operation, thereby avoiding the need of separate division. Proposed watermark embedding algorithm involves integer arithmetic, which is suitable for Integer DCT. The proposed scheme outperforms other schemes because they involve floating point arithmetic, which increases the complexity, thereby degrading real-time performance.

Step 6: The predicted values AC' are added with Δ to generate new AC values. Δ is a user defined constant ranging from 1 to 5.

If watermark bit $= 1$ then,

$$AC_i \geq AC_i' + \Delta$$

Else

$$AC_i < AC_i' - \Delta \tag{2.10}$$

Step 7: After embedding the watermark, inverse Integer DCT is performed to get a watermarked frame. These individual frames are assembled to generate the secure video.

The overall steps of watermark embedding are briefly summarized in Figure 2.3.

2.3.2 Watermark Extraction

Input: Watermarked video

Output: Extracted watermark image

The detection of proposed algorithm is blind wherein the original video is not required. The required steps for extraction are depicted in Figure 2.4.

Step 1: Secure video is divided in different groups of frames according to scene change detection method.

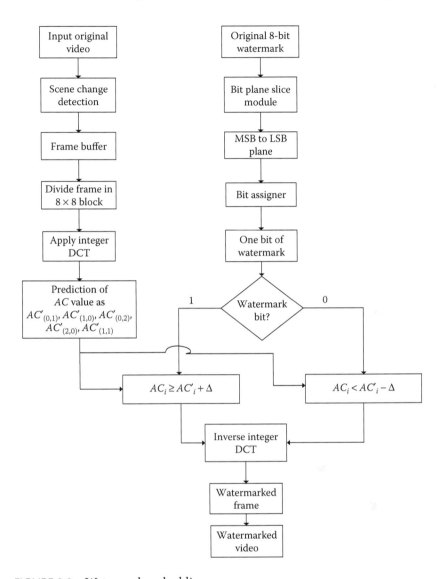

FIGURE 2.3   Watermark embedding process.

Step 2: Each frame of the scene is divided into 8 × 8 blocks to apply Integer DCT.

Step 3: The estimation of AC value is obtained with the help of the DC values of the surrounding blocks as given in Equations 2.5 through 2.9, which were also used in the above-mentioned embedding process.

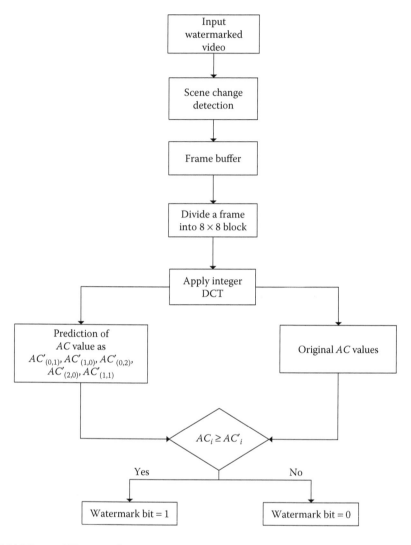

FIGURE 2.4 Watermark extraction process.

Step 4: The watermark bit is obtained by comparing the original value of AC with estimated value AC'.

For, $AC \geq AC'$

Watermark bit $= 1$

Else,

$$\text{Watermark bit} = 0 \qquad (2.11)$$

Step 5: The different bit planes are generated from the watermark bit.

Step 6: The bit planes are then assembled to generate the watermark image.

2.4 IMPLEMENTATION OF PROPOSED VIDEO WATERMARKING

The proposed video watermarking consists of two units: first is the watermark insertion unit and second is the watermark generation unit. These units work independently of each other and fetch data at the same time, thereby improving the performance of the proposed real-time video watermark embedding. Watermark insertion unit has two steps: first Integer DCT is used for spatial domain to frequency transformation, and the second unit is a prediction module, which is used to estimate the AC values from surrounding DC values. VLSI architecture of proposed watermarking is designed with pipeline and parallel data paths as shown in Figure 2.5.

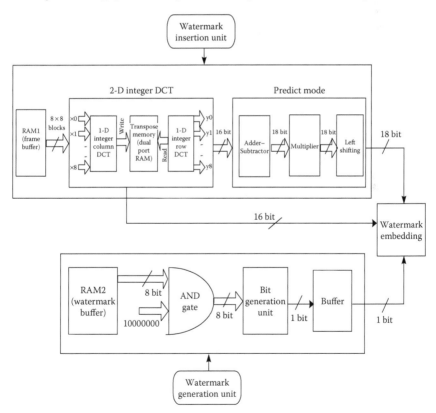

FIGURE 2.5   Proposed video watermarking architecture.

2.4.1 Watermark Insertion Unit

The coefficients of each frame are stored in RAM1 for watermark insertion unit. The watermark insertion unit has the first module of Integer DCT, where 8 × 8 data are applied to have 2-D DCT transformed value at every clock cycle. The architecture is designed for 2-D DCT, which has separable property, where the whole 2-D DCT is divided in two 1-D DCT processes. First, column DCT is calculated and then row DCT is computed in the same manner. The output of column DCT is stored in transposed memory. The dual port RAM is used as transpose memory where the data are written column wise and then read row wise for the second 1-D DCT process. Each 1-D Integer DCT is designed with fast butterfly structure to compute transformed values.

At every clock, eight inputs of 8-bit data width are fetched from RAM1. Integer 1-D DCT generates eight outputs of 12-bit data width after a clock cycle delay. The calculated column integer 1-D DCT values are stored in transpose memory and are applied for further row processing, which generates 16-bit width output of the 2-D Integer DCT coefficient. The output of 2-D DCT is then used for prediction of AC values. The predicted values are generated with adder/subtraction, multiplication, and shifter after one clock delay. They are compared with original values and then Δ (delta) is added according to a watermark bit.

2.4.2 Watermark Generation Unit

The main function of this unit is to generate one bit of watermark. Eight-bit watermark image is used as watermark and is stored in RAM2. Bitwise AND operation is carried out between the watermark image and a previously stored value (1000 0000). The output is assigned to the bit generation unit where each 8-bit data are compared with 0. If the value is greater than 0 then watermark bit is 1, else the bit is assigned to 0.

2.5 RESULTS AND ANALYSIS

The algorithm is tested against various attacks and analyzed on MATLAB platform. To check the robustness of the algorithm, the correlation parameter is verified with similarity factor (SF). It is calculated as follows:

$$\mathrm{SF}\left(w,w'\right)=\frac{\sum_{i=1}^{N-1}w\times w'}{\sqrt{\sum_{i=1}^{N-1}w^2}\sqrt{\sum_{i=1}^{N-1}w'^2}} \tag{2.12}$$

where:

W is the original watermark

W' is the extracted watermark

N is the size of watermark

For two identical frames, SF should be 1. To prove the authenticity of the video, the retrieved watermark should be similar to original watermark. The acceptable value of SF is about 0.85 (85%). Table 2.1 shows the performance of the algorithm in terms of SF against various temporal attacks. The performance of the algorithm is verified from the excellent results of experimental simulation. The algorithm uses scene change detection concept to mitigate the effect of temporal attacks.

The watermark gets distributed among different frames of the scene so the corruption of a particular frame would not lead to the loss of the entire watermark. Any part of the watermark can be recovered from the existing frames. If a frame gets dropped or corrupted then the total frame count is reduced. In such cases, the watermark is retrieved from average values of eight subsequent frames. The algorithm is also checked against some standard attacks that are applicable to a particular frame and results are elucidated in Table 2.2.

The watermark should be inserted to satisfy invisibility criteria for the observer. The visibility of the watermarked frame is checked with parameters defined as peak signal to noise ratio (PSNR) and mean square error (MSE).

$$\text{MSE} = \frac{1}{M \times N} \sum_{i=0}^{M-1} \sum_{j=0}^{N-1} \left[I(m,n) - I_w(m,n) \right]^2 \qquad (2.13)$$

TABLE 2.1 Performance of Proposed Algorithm against Various Temporal Attacks

Attacks	Percentage	SF
Frame dropping	10	0.9834
	20	0.9784
	30	0.8741
Frame swapping	10	0.9992
	20	0.9913
	30	0.9834
Frame averaging	10	0.9967
	20	0.9672
	30	0.9308

TABLE 2.2   Performance of Algorithm against Standard Attacks

Attacks	MSE	PSNR	SF
Gaussian noise (0 mean and variance 0.01)	10.5921	31.8116	0.9423
Salt and pepper noise (density 0.02)	15.2537	28.1644	0.9021
Compression (10%)	5.2300	38.8685	0.9789

$$PSNR = 20\log_{10} \frac{255}{MSE} \tag{2.14}$$

where:

   $I(m, n)$ is the original frame
   $I_w(m, n)$ is the watermarked frame
   M, N is the size of the frame

In Equation 2.13, MSE defines the number of error in bits that is generated between original and watermarked frame. The watermark embedding process changes the value of the coefficient, which leads to introduction of error between original and watermarked frame. But, this error should be perceptually invisible to the observer. From Equation 2.14, higher values of PSNR imply the lesser error encountered. In Table 2.2, MSE and PSNR values suggest the amount of error introduced because of an attack on the frame. The algorithm is also capable of sustaining most of the attacks and is still able to retrieve the watermark.

   Akiyo.yuv (frame size of 720 × 486) is considered as an original test video and 8-bit Lena image (64 × 64) is used as the watermark; the first frame of the video and the watermark are shown in Figure 2.6a and 2.6b, respectively. The resultant frame of the watermarked video is shown in Figure 2.6c. Similarity between the original and extracted watermark is 1 in the absence of any attack. Average PSNR and MSE values of original and watermarked video are

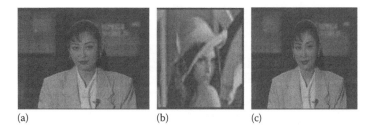

(a)   (b)   (c)

FIGURE 2.6   Simulation results (a) original video, (b) watermark, and (c) watermarked video.

found to be 69.19 and 0.252, respectively. The obtained values suggest that the proposed algorithm satisfies invisibility criterion during watermark insertion.

The proposed video watermarking technique is prototyped on FPGA to verify the real time performance for customized hardware. The algorithm is implemented on SPARTAN 3A DSP FPGA platform using Xilinx ISE 11.1. The resultant device utilization and resource utilization are reported in Tables 2.3 and 2.4, respectively.

The timing analysis of proposed algorithm is in Table 2.5. This showcases the simplicity of the proposed scheme. Thus, the algorithm can be easily integrated in H.246 for real-time watermark embedding.

The various modules of the proposed video watermarking scheme are synthesized with Design Compiler tool of Synopsis for 0.18 μm technology. The resultant area and power consumed are tabulated in Table 2.6. Integer

TABLE 2.3   FPGA Report for Resource Utilization

Resources	Watermark Generation	Watermark Insertion
Basic elements	3423	714
Registers	679	221
Adder–Subtractor	56	10
Multiplier	0	3

TABLE 2.4   Report for Device Utilization

Resources	Watermark Generation	Watermark Insertion
Slice	1832 out of 16840 (11%)	619 out of 16840 (3%)
Slice FFs	451 out of 33280 (1%)	167 out of 16840 (0%)
4 input LUTs	3259 out of 33280 (9%)	1299 out of 33280 (3%)
Bounded IOBs	131 out of 519 (25%)	60 out of 519 (11%)
Global clocks	1 out of 24 (4%)	1 out of 24 (4%)

TABLE 2.5   Synthesis Report for Timing Analysis

Parameter	Watermark Generation	Watermark Insertion
Frequency (MHz)	37.551	30.68
Minimum period (nS)	26.63	32.59

TABLE 2.6   Synthesis Results for Proposed Watermark Algorithm

Block	Module	Area (μm²)	Power (mW)
Watermark insertion unit	1-D DCT	16608.00	117.30
	2-D DCT	24064.00	197.36
	Prediction module	951.00	6.6475
Watermark generation unit	Bit plane slice	192.00	0.023

DCT is obtained with two 1-D DCT operations for row and column, and the area of each stage is listed in Table 2.6. The results for watermark insertion and generation units are also mentioned.

A comparison of proposed algorithm with existing video watermarking custom IC implementations is presented in Table 2.7. Most of the previous works (Strycker et al. 2000; Maes et al. 2000; Mathai et al. 2003a; Tsai and Wu 2003) were based on spatial domains, which have uncompressed domain watermarking. All these watermark embedding algorithms require partial encoding/decoding hardware in order to embed the watermark for the chosen compression standard. Due to this shortcoming, they are not useful for real-time operation. Previous work by Vural et al. (2005) designed watermarking method for wavelet-based frequency domain but failed to fit as part of encoder because all the standards have DCT for transformation. The article by Mohanty et al. (2009) suggests perceptual watermarking architecture for MPEG-4 and requires complex computation, which restricts its usage for H.264 standard. Another article (Roy et al. 2011) designed efficient invisible and semi-fragile video watermarking algorithm, which can be integrated with MJPEG video compressor unit. This algorithm uses general DCT-based watermarking that has floating point structure. The proposed algorithm provides comparable results in terms of speed, power, and area. The designed architecture fulfils the real-time requirement for watermark embedding criteria. This algorithm is the only attempt for video watermarking for the latest H.264 standard. The algorithm provides better speed due to the involvement of integer arithmetic.

2.6 CONCLUSIONS

This chapter presented an invisible and robust watermarking scheme suited for H.264-based video codec. The algorithm was synthesized using Faraday 0.18 μm CMOS technology and is implemented on SPARTAN 3A FPGA platform. Integer DCT was calculated with separable property, which allowed 1-D Integer DCT calculation for row and column separately. The proposed video watermark has parallel and pipeline data path to increase speed for watermark embedding process. The speed of video watermarking unit was found to be sufficient for real-time requirements. The proposed video watermarking exhibits excellent robustness against different types of temporal attacks. The area and power consumption of the proposed algorithm suggest that custom IC of proposed scheme can be

TABLE 2.7 Comparison with Previous Work

References	Watermark Type	Domain	Standard	Frame Size	Technology	Area	Power (mW)	Speed (MHz)
Strycker et al. (2000)	Invisible robust	Spatial	Raw video	144 × 360	NA	NA	NA	100
Maes et al. (2000)	Invisible robust	Spatial	Raw video	NA	NA	17 kG	NA	NA
Mathai et al. (2003b)	Invisible robust	Spatial	Raw video	320 × 320	0.18 μm	3.53 mm^2	60	75
Tsai and Wu (2003)	Invisible robust	Spatial	Raw video	NA	NA	NA	NA	NA
Vural et al. (2005)	Invisible robust	DWT	MPEG	512 × 512	NA	NA	NA	NA
Mohanty et al. (2009)	Visible	DCT	MPEG-4	320 × 240 × 3	NA	NA	NA	100
Roy et al. (2011)	Invisible semi-fragile	DCT	MJPEG	640 × 480	0.18 μm	NA	10	40
Proposed	Invisible robust	Integer DCT	H.264	720 × 486	0.18 μm	5.25 mm^2	104.80	120

integrated with a standard H.264 encoder to generate a standard secured video bit stream.

ACKNOWLEDGMENTS

The authors thank Prof. Anand Darji, coordinator of the SMDP-II Lab, Sardar Vallabhbhai National Institute of Technology (SVNIT), Surat, for his support. Thanks are also due to the Ministry of Human Resource Development, Government of India, for funding the SMDP-II Lab, which has provided facilities to use EDA tools like design compiler (Synopsys), ModelSim (Mentor Graphics), and Xilinx ISE. The authors are also thankful to each and every one who has been associated directly or indirectly with this work.

REFERENCES

Dengpan, Y., Zhuo, W., Xuhua, D., and Deng, R.H. (2012), Scalable content authentication in H.264/SVC videos using perceptual hashing based on Dempster-Shafer theory, *International Journal of Computational Intelligence Systems*, Vol. 5, issue 5, 953–963.

Jayanthi, V.E., Mani, R.V., and Karthikeyan, P. (2012), High performance VLSI architecture for block based visible image watermarking, *International Journal of Electronics*, Vol. 99, issue 2, 1191–1206.

Kougianos, E., Mohanty, S.P., and Mahapatra, R.N. (2009), Hardware assisted watermarking for multimedia, *Special Issue on Circuits and Systems for Real-Time and Copyright Protection of Multimedia, Computers Electrical Engineering*, Vol. 35, issue 2, 339–358.

Lee, M.-S. (2003), Image compression and watermarking by wavelet localization, *International Journal of Computer Mathematics*, Vol. 80, issue 4, 401–412.

Maes, M., Kalker, T., Linnartz, J.P.M.G., Talstra, J., Depovere, G.F.G., and Haitsma, J. (2000), Digital watermarking for DVD video copyright protection, *IEEE Signal Processing Magazine*, September, 47–57.

Malvar, H.S., Hallapuro, A., Karczewicz, M., and Kerofsky, L. (2003), Low-complexity transform and quantization in H.264/AVC, *IEEE Transactions on Circuits and Systems for Video Technology*, Vol. 13, issue 7, 598–603.

Mathai, N.J., Kundur, D., and Sheikholeslami, A. (2003a), Hardware implementation perspectives of digital video watermarking algorithms, *IEEE Transactions on Signal Processing*, Vol. 51, issue 4, 925–938.

Mathai, N.J., Sheikholeslami, A., and Kundur, D. (2003b), VLSI implementation of a real-time video watermark embedder and detector, *Proceedings of the IEEE International Symposium on Circuits and Systems*, Vol. 2, 772–775.

Mohanty, S.P., Kougianos, E., Wei Cai, and Ratnani, M. (2009), VLSI architectures of perceptual based video watermarking for real-time copyright protection, *Proceedings of the 10th International Symposium on Quality Electronics Design*, 527–535.

Piva, A., Bartolini, F., and Barni, M. (2002), Managing copyright in open networks, *IEEE Transactions on Internet Computing*, Vol. 6, issue 3, 18–26.

Richardson, I.E. (2004), *H.264 and MPEG-4 Video Compression (Video coding for next generation multimedia)*, John Wiley & Sons, England.

Richardson, I.E. (2010), 4 × 4 transform and quantization in H.264/AVC, *White Paper on Video Compression Design, Analysis Consulting and Research*, Vcodex, 2010.

Roy, S.D., Xin Li, Yonthan, S., Alexander, F., and Orly, Y.P. (2011), Hardware implementation of a digital watermarking system for video authentication, *IEEE Transactions on Circuits and Systems for Video Technology*, Vol. 23, issue 2, 289–301.

Strycker, L.D., Termont, P., Vandewege, J., Haitsma, J., Kalker, A., Maes, M., and Depovere, G. (2000), Implementation of a real-time digital watermarking process for broadcast monitoring on trimedia VLIW processor, *IEE Proceedings on Vision, Image and Signal Processing*, Vol. 147, issue 4, 371–376.

Tsai, T.H., and Wu, C.Y. (2003), An implementation of configurable digital watermarking systems in MPEG Video Encoder, *Proceedings of the IEEE International Conference on Consumer Electronics*, 216–217.

Voloshynovskiy, S., Pereira, S., Pun, T., Eggers, J.J., and Su, J.K. (2001), Attacks on digital watermarks: Classification, estimation based attacks and benchmarks, *IEEE Communications Magazine*, Vol. 39, issue 8, 118–126.

Vural, S., Tomii, H., and Yamauchi, H. (2005), Video watermarking for digital cinema contents, *Proceedings of the 13th European Signal Processing Conference*, 303–307.

Wiegand, T., Sullivan, G. J., Bjontegaard, G., and Luthra, A. (2003), Overview of the H.264/AVC video coding standard, *IEEE Transaction on Circuits and Systems for Video Technology*, Vol. 13, issue 7, 560–576.

CHAPTER 3

FPGA-Accelerated Algorithm for the Regular Expressions Matching System

Paweł Russek and Kazimierz Wiatr

CONTENTS

3.1 INTRODUCTION

This chapter describes an algorithm to support a regular expressions matching system. The goal was to achieve an attractive performance system with low energy consumption. The basic idea of the algorithm comes from a concept of the Bloom filter. It starts with the extraction of static substrings for strings of regular expressions. The algorithm is devised to gain from its decomposition into parts that are intended to be executed by custom hardware and the central processing unit (CPU). The pipelined custom processor architecture is proposed, and a software algorithm explained accordingly. The software part of the algorithm was coded in C and runs on a processor from the ARM family. The hardware architecture was described in Very High Speed Integrated Circuit (VHSIC) hardware description language (VHDL) and implemented in field programmable gate arrays (FPGAs). The performance results and required resources of the above experiments are given. An example target application for the presented solution is computer and network security systems. The idea was tested on nearly 100,000 body-based viruses from the ClamAV virus database. The solution is intended for the emerging technology of clusters of low-energy computing nodes.

3.2 MOTIVATION

Regular expression matching has become a bottleneck of software-based pattern matching systems. Such systems play an important role in solutions detecting computer viruses, malware, and other malicious threats. A class of these systems is network intrusion detection system (NIDS). The main components of NIDS are packet classification and deep packet inspection (DPI). In practice, traditional firewalls perform packet classification. They check specific fields of the packet header. The DPI is used to recognize and detect network attacks, especially in the application layer, that general firewalls cannot find. Such systems perform scanning and filtering of the network, checking over the application data layer by matching the data against thousands of signatures or strings that belong to previously known attacks. The signature databases are frequently updated with new emerging traffic features. For example, Snort [1], Bro [2], and Linux L7-filter [3] offer the rules sets. These rules can be used to extract patterns that DPI systems later match against packet payloads. In this chapter, ClamAV [4], an open source antivirus toolkit, will be inspected closer.

Data matching tasks may consume a lot of CPU runtime in software systems. As the number of signatures has recently increased dramatically, matching requires more and more computational effort. ClamAV contains nearly 100,000 body-based patterns as of January 2013.

On the other hand, pattern matching is an example of data-intensive computing. In other words, it is not a CPU-bound problem, but rather input/output (IO) bound problem. For such problems, the computer architecture should provide efficient data movement in the system. Present state-of-the-art, superscalar, many-core, out-of-order execution processors fall behind real needs for data-intensive problems. For example, modern 64-bit processors offer over 200 billions of floating-point operations per second (GFlops) of processing power. When considering a $O(n)$ complex problem then to accommodate such performance, data transfer of 1.6 trillion bytes per second (TBps) is requested. The comparison of that number to available network traffic speed shows a tremendous gap. Additionally, the cost and the energy consumption of powerful processors need to be considered.

A possible approach to solve the above problems is a computer system that is better balanced for data-dependent computing. Such systems can be built around low-energy processors. They offer low computing power but feature attractive ratio of performance per consumed energy. What is most important, their computing power is high enough to fit available IO data throughput for some class of algorithms. Low-energy processors based systems compensate the lower CPU's performance with a higher number of computing nodes' use for a task. Thus, such a task is executed on many computing nodes. It is very important that by adding an additional node, one adds CPU performance together with an additional data throughput introduced by its IO devices. To give an example of such a system, HP's Moonshot solution can be taken into consideration [5]. The target applications for low-energy servers are usually high-scale Internet services. Fetching and delivering data is more important than computational power in such systems.

Another approach that plays a significant role in data processing is custom processing. In terms of performance and energy consumption, custom processors outperform general-purpose processors. Today, field-programmable gate array (FPGA) is a proven semiconductor technology that is used for implementation of custom computing devices for data-intensive algorithms. Examples can be found in Jamro et al. [6] and Russek and Wiatr [7].

The use of dedicated processor architecture is advised for algorithms that exhibit simple data flow and can be trivially parallelized or pipelined. This means in practice that only selected parts of real-life applications can be successfully implemented in hardware. A successful implementation is the one that introduces an improvement in terms of performance and/or energy consumption when compared to the CPU. This is the reason that only so-called computational kernels are executed in hardware. The majority of application code remains as a software code to be executed by a CPU. This leads us to the concept of the popular hardware–software (HW–SW) codesign approach.

Also, a recent introduction of Xilinx's Zynq [8] is the motivation to focus research on capabilities of low-energy processors, which are tightly coupled with an FPGA structure. The Zynq-7000 Extensible Processing Platform is a family of silicon devices that combine a complete ARM processor-based system-on-chip (SoC) with integrated programmable logic. Each device is capable of booting an OS from reset with the programmable logic accessible. It enables system designers to apply to an application execution a combination of software (using the ARM processor) and hardware (using programmable logic) processing.

In the presented work, the problem of regular expression matching is approached in the manner of HW–SW codesign. The algorithm was proposed, and its performance tested on the Zynq-7000 platform. We applied the bottom-up approach as the algorithm was conceived from scratch with target platform architecture kept in mind. Another approach, to migrate existing algorithm to a new hardware platform is more frequent, but it was not applied here because, according to the authors' experience, it usually delivers worse results.

The chapter is organized as follows. The next section presents recent and related works that were chosen from literature. In Section 3.4 an outline of issues that are the background of the work is given. Section 3.5 presents a decomposition of the problem into hardware and software parts. Also, principles of a virus database decomposition are given. A complete algorithm idea is introduced in Section 3.6. In Section 3.7, the architecture of the proposed custom hardware accelerator is presented. Results of hardware and software implementation are discussed in Section 3.8. The chapter ends with conclusions in Section 3.9.

3.3 RELATED WORK

The rapid increase of network attacks and data traffic show that traditional software-only NIDS may be ineffective. Many solutions have been proposed to overcome this problem. In contrast to software-only regular expressions systems, many studies proposed hardware architectures for accelerating attack detection. Deterministic finite-state automata (DFA) and non-deterministic finite-state automata (NFA) algorithms are the two main approaches that are used to perform DPI by regular expression matching. NFAs are smaller in size since the number of states in an NFA is usually comparable to the number of characters presented in its regular expressions. On the other hand, DFA works faster by proceeding one character each clock cycle, but the DFA has the potential state explosion problem for complex regular expressions.

Hardware architectures using an FPGA became popular for NIDS implementations because the FPGAs allow updating of new attack patterns. A good example of successful FPGA-based NIDS system was given by Clark et al. [9]. One of the main challenges of hardware implementation is to accommodate a large number of regular expressions to FPGAs. The minimization of requirements for logic resources has been studied intensively. DFA representations are typically used to implement regular expressions at high network link rates. Kumar et al. [10] introduces a new representation for regular expressions called the *delayed input DFA* (D2FA), which reduced space requirements as compared with the DFA. A D2FA is constructed by transforming a DFA via incrementally replacing several transitions of the automaton with a single default transition. The implemented system required less than 2 MB of embedded memory and provided up to 10 Gbps throughput at a modest clock rate of 300 MHz. The set of regular expressions contained approximately 750 moderately complex expressions, and Xilinx Virtex-4 FPGA family was used.

Sidhu and Prasanna [11] proposed to construct an NFA from a regular expression to perform string matching. Later, Hutchings et al. [12] presented a tool to generate an NFA that shared common prefixes to spare the FPGA resources. Following further improvements were proposed. The latest achievements in sharing common subregular expressions for the NFA

implementaton can be found in Lin et al. [13]. The total number of implemented Snort rules was 24,214 characters. The reported throughput was
approximately 1 Gbps. Sourdis et al. [14] improved gained rates by introducing new basic building blocks to support constraint repetitions syntaxes
more efficiently than previous works. The design methodology was supported by a tool that automatically generated the HDL code for the given
regular expressions. Thanks to parallelism in multicharacter architecture,
Chang et al. [15] were able to increase the maximum system throughput to
4.68 Gbps (2-character design) and 7.27 Gbps (4-character design).

A clear limitation is the size of the signature database that could be
compiled into an FPGA. To alleviate that problem, Yang et al. [16] introduced an architecture that uses the FPGA's BRAM modules. The complex
character classes were matched by a BRAM-based classifier shared across
different regular expressions. The authors reported a 10 Gbps throughput.
The BRAM-based approach was continued by Wanga et al. [17].

Bispo and Cardoso [18] presented the synthesis of regular expressions
with the aim of achieving high-performance engines for FPGAs. They
proposed new solutions for constraint repetitions and overlapped matching. As a case study, they presented FPGA implementations for two rule-
sets of Bleeding Edge and Snort.

Bande Serrano and Palancar [19] proposed an approach that is related
to and presented in our work. Unique subsequence matchings were used
for detecting the possible presence and the corresponding alignment of
the string in the data flow. In doing so, they made a reduction of the area
cost for processing multiples characters.

For the purpose of regular expression matching, other approaches that
are related to finite automata were also presented. Bando et al. [20] introduced lookahead finite automata (LaFA). For LaFA, 34 Gbps estimated
throughput could be achieved, and the FPGA chip could accommodate
up to 25,000 regular expressions. Wanga et al. [17], using instruction
finite automaton (IFA), demonstrated Snorts 1083 regular expressions and
11.3 Gbps rate for single Altera Stratix IV EP4SGX530NF45C2ES.

Bloom filtering [21] is another essential algorithm used for pattern matching. Its hashing and look-up technique lead to very fast implementations. It
also has very low resource requirements. The Bloom filter is a choice for real-
time, embedded applications. For example, Jamro et al. [6] implemented
several parallel Bloom filters in a single FPGA structure. As a result of three-
level parallelism and a 100 MHz clock frequency, a theoretical throughput
of 12.8 Gbps was achieved. Although the original Bloom's algorithm is

intended for static patterns, several authors tried its adoption for regular expression matching. A more recent approach was given by Dharmapurikar et al. [22]. The authors proposed a hashing table look-up mechanism utilizing parallel Bloom filters to enable a large number of fixed-length strings to be scanned in hardware. Based on the Bloom filter, Lockwood et al. [23] proposed the gateway that provides the Internet worm and virus protection in networks. In Dharmapurikar et al. [22] and Dharmapurikar and Lockwood [24], the Bloom filter, instead of including just present bits at hash locations, included an address for the microcontroller. Some regular expressions could be supported by linking string literals together with software.

A few studies have used the ClamAV pattern set in their evaluations. Ho and Lemieux [25] and Tsung Lin Ho and Lemieux [26] used Bloomier filters in their PERG and PERG-RX architectures. The Bloomier filter is an extension of the Bloom filter. PERG supports the single-byte and displacement wildcards ? and {n}, which insert fixed-length gaps between string fragments. PERG-RX adds support for other wildcards that require arbitrary-length gaps. It operates at a rate of 1.2 Gbps.

3.4 BACKGROUND

To let a computing system run four elements are necessary: an algorithm, data structure, hardware, and software. In this section, important topics related to a computing platform will be highlighted first. Then, basic virus database concepts will be introduced. Finally, the Bloom filter algorithm will be introduced.

3.4.1 Computing Platform

A processor requires an appropriate host system architecture. In a well-balanced system, performance of all components should fit. However, in the case of general-purpose processor optimized properties of a host system cannot be strictly defined, that is, they are different for various algorithms. In practice, the best general-purpose solutions are built using state-of-the-art components: processors, memory chips, graphics card, storage devices, and so on. The policy of choosing the best available component on the market to set up a computer works in practice as such systems execute different kinds of applications. But, for IO bound tasks, cutting-edge processors do not help when the storage is too slow. Conversely, when a single-purpose system is proposed, it is possible to establish a proper balance between computer components' performances.

An essential computing system equipped with a hardware accelerator is depicted in Figure 3.1.

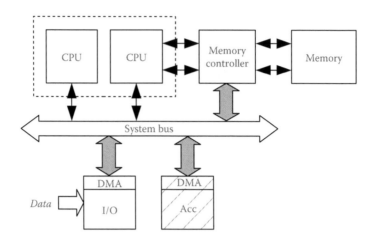

FIGURE 3.1   Simple system with an accelerator.

It consists of a multi-CPU processor, a memory controller connected to a memory block, an IO device, and a hardware accelerator. The CPU is tightly coupled with the memory controller that allows for higher data throughput and lower latency. In contrary, the IO devices and the accelerators have access to the memory, thanks to a system bus (e.g., PCIe or AMBA bus). As the memory controller is also connected to the system bus, the IOs and accelerators can also perform memory operations. Preferably, the system bus devices use direct memory access (DMA) transactions to access the main memory. The DMA mechanism allows the CPUs, IO devices, and accelerators to work simultaneously. In principle, they perform algorithms in a pipeline manner. The functional decomposition of an algorithm is done, that is, separate blocks of data are processed consequently by the IO device, CPU, accelerator, and so on. The performance of all components must fit to allow a smooth algorithm execution. This policy means that no component outperforms other components, and no component waits for other components to complete their task.

In the case of NIDS, the packet payloads in continuous packets are original content's fragments and may be out of order. Packet reassembly reconstructs the correct application data at the end host. Typical speeds reported for existing packet reassembly systems are 1 to 10 Gbps under normal traffic, which is the performance requirement for NIDS components.

3.4.2 Viruses Database

A typical database of computer viruses stores so-called signatures of the viruses' binary execution codes. A signature is a selected fragment of a corresponding virus code. Depending on the type of the signature, it can be stored in various formats. For a static malware, a checksum can be applied. In that case instead of a complete binary representation, only the appropriate checksum is backed up. A basic method that can be used to detect such patterns is the Bloom filter. The Bloom filter can be implemented either in hardware or in software. Checksums can be treated as Bloom's hashes. Static virus codes are not the interest of this chapter. Another class of signatures in the database is the body-based signatures class. Regular expressions are used to describe those signatures. Those expressions represent binary viruses' footprints that are strings that can be additionally enhanced using wildcards. There are various types of wildcards. The meaning of a wildcard can be *match any byte, match any byte string, match any of selected bytes, don't match selected bytes*, and so on. That format resembles syntax of regular expressions that are commonly used for string matching. It allows a storage to keep an array of virus mutations as a single pattern definition. In general, the viruses that are expressed in the second form are more cumbersome to detect by an antivirus system. This chapter addresses body-based virus signatures only.

3.4.2.1 ClamAV Signatures

ClamAV is an open source (GPL) antivirus engine designed for detecting Trojans, viruses, and other malware. ClamAV virus database contains over 1 million virus signatures. ClamAV signatures are stored in the database in a defined format. There are two ways of storing malware specific signatures. The easiest way is to store MD5 checksums of dangerous binary chunks, but this method applies to static virus patterns only. The more sophisticated and flexible method is a body-based method. This method implements the concept of wildcards. Bodies of the viruses are stored in a database as is, except that wildcards are used to cover different mutations of the virus. In principle there are two types of wildcards in ClamAV: a single character related wildcards and wildcards that are related to a string of random length.

The character wildcards replace a single character in a signature chunk. They can be defined as a set of allowed characters. They are expressed in a form (*aa*|*bb*|*cc*|...), which means to match any byte from a set of hexadecimal values: {*aa*, *bb*, *cc*}. An alternative form for character wildcards are: ?*a*, *a*?, and ??, where the symbol ? denotes *a random nibble* (a nibble is 4 bits). The string wildcards represent a random string within a signature body. The length of the string can be: within a defined range for wildcards in a form of {*n* − *m*}, *less than* for the form {−*n*} or *greater than* for the form {*n*−}.

Here, in this chapter, character wildcards are denoted as *?* and string wildcards are denoted as ***. A part of a signature that is located between any *** wildcards is called in the chapter signature subchunks. The important feature is that the signature subchunks have a static length.

3.4.3 Bloom Filter

The fundamental operations that are behind the Bloom filter are both data hashing and memory read. In general, the hashing converts a word into another word of a smaller bit length. In other words, the hashing process converts a w-bit input word into an h-bit output word, where $h < w$. The output word that is generated by the hashing process is treated as an address in a memory where the stored values are *true* or *false*. These indicate the existence or nonexistence of the input word in the filter. In the teaching phase, the memory is programmed for the given word dictionary. In the check phase, if the location in the memory at the generated address is *true*, a hit occurs. In Bloom's original algorithm, for a single word, the mentioned memory operation is sequentially repeated several times for different hash functions. We denote this number of tries as k. If all the tries are successful, during the check phase, a positive match is generated, that is, the word exists in the Bloom filter.

There are at least two major disadvantages of the Bloom filter method that need to be considered here. First, the Bloom algorithm is not a trustworthy tool. Second, in the original algorithm, the search is performed sequentially.

About the first disadvantage, there is a finite probability of an incorrect word match. The Bloom algorithm generates a match signal for the words outside the dictionary. They are further denoted as *false-positives*. If fault-lessness is required, the results ought to be additionally verified by another, perfect algorithm. The nature of the fault indicated in the Bloom filter is a misstatement, that is, the words that are not in the dictionary can match. Fortunately, the opposite situation, where the algorithm overlooks words is not possible. This means that the words held in the dictionary are always indicated.

The probability of *false-positives* is given by Equation 3.1.

$$P_{\text{err}} = \left[1 - \left(1 - \frac{1}{m} \right)^{(kn)} \right]^{k} \approx \left[1 - e^{-kn/m} \right]^{k} \qquad (3.1)$$

where:

    n is the number of searched patterns

    m is the bit size of memory

    k is a number of hash operations performed

Additionally, a reasonable assumption is that $m = 2^h$.

3.5 PRINCIPLES OF A PROPOSED ALGORITHM

In this section, the basic idea of the proposed algorithm will be introduced. The main assumption that is behind the proposed scheme is its hardware acceleration and real-time execution. That is why a part of the algorithm will be executed by the custom processor. The streaming architecture of the accelerator was chosen as it offers the highest performance. Additionally, data stream processing model naturally suits network data processing.

It was already mentioned that a method that can be very efficiently implemented in hardware is the Bloom filter and its variations. However, the Bloom filter does not allow for regular expressions matching. On the other hand, the advantages of hashing cannot be neglected so in order to gain performance, our algorithm uses hashing. As a consequence, to get rid of *false-positives*, the hashing results must be verified. Additionally, to cover regular expression matching our algorithm introduces extended verification.

3.5.1 Hardware/Software Decomposition

Obviously, any method that inherits from the hashing approach cannot be applied directly for detection of viruses that use a format of regular expressions. In our method, hashing is applied to static fragments of viruses' bodies only. Those fragments are called virus traits. It is possible to search for the traits of the viruses and when a match occurs to conduct regular expression comparison.

Accordingly, our malware detection approach consists of two phases. In the first phase, data is examined against the virus's traits, and then suspected blocks of data are passed to the second phase of deeper examination. Regular expression matching is implemented in a second phase.

In comparison to the Bloom filter, the proposed hashing-based scheme offers an additional advantage. Here, the first phase additionally generates information containing the reference to a certain virus that needs to be checked. Thanks to this, the second phase is sped up as it is applied to a selected database's viruses only. This method will be further explained in Section 3.7.1.

In the proposed scheme, the first phase is executed in hardware and the second phase in software.

3.5.2 Database Preprocessing

The body-based signatures must be processed first to apply the above method. In the first stage, traits $T(i) \in T$ of arbitrary length t have to be selected for each virus entry $V(i) \in V$ ($0 < i < n$, where n is a number of viruses in the database). The trait $T(i)$ is a static substring of a virus body definition, so it does not contain any wild cards. In perfect circumstances, each virus body $V(i)$ has a unique trait $T(i)$. The value t should be big enough to allow it to distinguish different $V(i)$ but sufficiently small to fit candidate substrings between wild cards in a virus body. It must be noted that each $V(i)$ has a different length v. The process of $T(i)$ selection for an arbitrary t can be performed by an exhaustive search algorithm. The pseudocode is given in Listing 3.1.

It is possible to repeat the above process for different values of t to achieve the satisfactory uniqueness of traits. In practice, for a very large number of viruses the values $T(i)$ will be duplicated for different signatures. All relevant signatures must be looked up during the verification step to cope with this. For the best system performance, we should find as many unique substrings as possible because the traits repetition for

LISTING 3.1　An Exhaustive Search for Viruses' Traits Extraction

```
function extract_traits (V, t, n)
for i in 1 to n
k=0 /*sub–strings iterator */
do
repeat=FALSE
T(i)= get_static_sub_string (V(i), k, t)
if (T(i)==NULL) break
for j in 1 to i
if (S(i)==S(j))
k=k+1
repeat=TRUE
break
while (repeat==TRUE)
/*end for*/
return T
```

different viruses can lead to an excessive number of check-up iteration in the verification phase.

In the next step of database preprocessing, strings $T(i)$ are compressed to hashes $H(i) \in H$. Hashes have a constant length h. The hashing process can be performed using any checksum function (e.g., CRCn, MD5, and SAX). It may not be clear why a hashing process is necessary. One can ask, why not generate the substrings $T(i)$ of length h at once. The reason is that when substrings are too short, their uniqueness may suffer. Because of virus body similarities, it is difficult to distinguish the viruses if their substrings are not long enough. Simply, sort sequences of code do not exhibit the differences in viruses. The effect of this phenomenon will be presented in example results given in Section 3.8.1.

3.6 5-STAGE ALGORITHM

For detecting body-based signatures, a *5-stage* algorithm is proposed. At each stage, it checks a single condition to confirm a pattern match. The pattern is fully verified when conditions at all stages are positively verified. The *5-stage* algorithm involves the following activities:

1. Signature trait detection

2. Trait verification

3. Main subchunk match

4. Forward subchunks search

5. Backward subchunks search

The algorithm flow is presented in Figure 3.2. Consecutive algorithm's stages are more and more computationally exhaustive. So stage s requires less time to complete than stage $s + 1$. The reason of that policy is that we want to reject the potential *false* pattern match as quickly as possible. This approach allows to release the CPU for other pattern verifications and potentially improves overall system performance. A distinct feature of the proposed scheme is that the hardware architecture will be introduced to execute the first stage of the algorithm.

Figure 3.3 explains the concept of a trait and subchunks that constitute a signature. For a given matching system, traits are static binary strings of arbitrary length t. Each trait is a specific part of a signature so that it contains no wildcards. Such a trait uniquely determines a corresponding

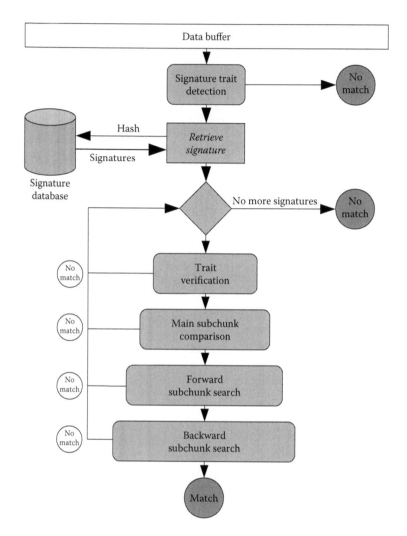

FIGURE 3.2   *5-stage* algorithm.

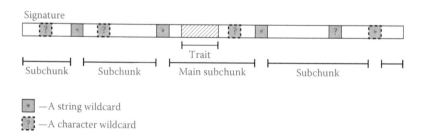

FIGURE 3.3   Composition of a virus signature.

signature and the signature body can be immediately retrieved from a database. Subchunks are signature's parts that lie between the adjacent type * wildcards. A subchunk can contain type ? wildcards in its body.

3.6.1 Signature Trait Detection Stage

Signature trait detection stage detects traits in an inspected data buffer. The hashing method can be used to implement its functionality. If *hit* occurs the stage returns a position P in the data buffer and a calculated hash value $H(i)$.

We propose the hardware implementation of this stage. The purpose of this stage is prefiltering of a binary input data stream. The rate of *hit* occurrences in this stage depends on the hash length h and the number of viruses n. To avoid a bottleneck in the algorithm flow, the h should be adjusted to fit the performance of a CPU, which executes consecutive algorithm's stages.

3.6.2 Trait Verification Stage

False-positives are possible at the first stage. The goal of trait verification stage is to check if the location that is returned by the previous stage contains a valid signature trait.

This stage and the stages that follow depend on an index that is used to address a signature database table. This index $X(i) \in X$ is an address of a record in an index table that contains a virus position in the database table. The position is used to retrieve an appropriate virus body quickly. The index value is derived from the hash $H(i)$ calculated at the first stage. The selected bits of the $H(i)$ are used to generate the value $X(i)$. The length of X is denoted as x. This technique is similar to *hash indexes*, which are utilized in database systems.

The single index can point to more than one signature in the database. This indicates that at single verification step, more than one signature must be checked. For better system performance, we prefer to limit the number of signatures that must be verified. This number can be adjusted by an x value. It is possible to approach a number of signatures that must be verified in a single step to one signature only. This is possible when $x > \lceil log_2(n) \rceil$. In practice, it may cause an index table to be prohibitively big to fit a system memory. That is some disadvantage of the used method.

Finally, a simple memory comparison function is used to compare data buffer at position P with the traits that were read from the database. This stage can be relatively quickly performed by the software.

3.6.3 Main Subchunk Match Stage

A subchunk that contains a trait is referred to as the main subchunk. If the previous stages positively confirmed existence of the trait in a buffer, the very next step is to match the main subchunk. This can be done relatively quickly. Knowing the trait position P, one can determine an expected location of the main subchunk in the data buffer. This stage performs a single buffer comparison. In contrast to a trait comparison, subchunks contain character wild cards ?. Thus, simple memory comparison must be preceded by wild card verification. Each wild card is verified independently. This makes the subchunks comparison more complex than the trait verification.

3.6.4 Forward and Backward Search Stages

The subchunks position within the data buffer is related to the trait position P returned by the first stage. Except the main subchunk, other subchunks can be located within the data buffer in many different positions. As subchunks in signature body are separated by string wildcards *, different positions in the data buffer must be checked for a certain subchunk existence. For example, a string wild card $\{n - m\}$ forces the algorithm to check the buffer in the relative position range from n to m. The forward subchunks search applies to subchunks that precede the main subchunk. The backward subchunks search identifies subchunks that follow. Forward and backward procedures are separated because when coded in programming language they use the different procedure functions. Those functions are iteratively called as long as the first and the last subchunks of the signature body are reached for the forward and backward stages, accordingly.

3.7 HARDWARE ACCELERATION

In order to achieve maximum system throughput, the 5-stage algorithm should be accelerated by a dedicated hardware coprocessor. As it was mentioned earlier in the chapter, the 5-stage algorithm working scheme is an execution of the consecutive verification stages, which reject the existence of the signature in the examined data. The signature confirmation is achieved by the successful execution of all the five stages. Additionally, the 5-stage algorithm philosophy is that the computational complexity of the algorithm grows as the higher number of a stage is performed. The policy behind such an approach is that the number of entries to each stage during the program execution should decrease as the verification proceeds.

As a consequence, the most computationally exhaustive part is the first stage of the algorithm. As mentioned before, this stage can be completed using the Bloom filter algorithm. However, because of the Bloom filter drawbacks another algorithm is incorporated in the first stage. The authors' method that derives from the Bloom's algorithm will be presented in Section 3.7.1.

3.7.1 Hash Binary Tree

The Hash Binary Tree (HBT) algorithm that is used for hardware implementation of the trait search stage will be presented in this section. The idea was derived from both search binary trees and the Bloom filter.

The binary tree is a well-known algorithm. It is presented in Cormen et al. [27], for example. The algorithm is used to find a match for data, in a set of reference values. The binary tree can be seen as a tree of consecutive decisions that are made by a binary search algorithm. The tree is depicted in Figure 3.4. The patterns from the sorted reference set must be properly located in the tree's nodes. The node $n(i)$ corresponds to ith value from the set. The input data element propagates from the tree stub to the top leaves. At each node element's value is compared to the reference value that is stored in the node. According to the results of the comparison, the data element continues downward to the left or right neighbor. The tree can be divided into levels. There are L levels in the tree. The levels are enumerated from 0 to $L - 1$. A set of reference data counts $2^L - 1$ elements.

To introduce a pipelining concept, we assume that at each algorithm's step, L, data elements can be processed in the tree simultaneously. In the

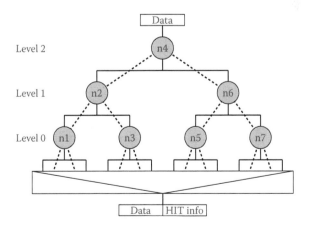

FIGURE 3.4 Binary tree organization.

hardware implementation of the tree, at every clock cycle, a single data element is processed on each tree level.

When match algorithms are implemented in hardware, a single comparison operation can be arbitrary bit-wide. In theory, we can compare patterns of any length in a single clock cycle. Of course, that length must be reasonable. The longer the bit-width, the higher will be the utilization of hardware resources. The above leads to the pattern compression idea. The same functionality of the tree can be achieved if the patterns are replaced by their unique hash representatives. This allows for hardware resources savings and leads to the HBT.

When we hash input patterns, we can treat HBT as a Bloom filter. The only difference is in the way the hash values are stored. In deference to a Bloom filter, HBT stores hash values in an explicit way. However, theoretical properties of the Bloom filter (including Equation 3.1) remain valid.

Dynamic pattern update is a functionality that is challenging for hardware matching. In the case of HBT, for the given dictionary of patterns the procedure of preparing an appropriate tree structure is very simple. It is enough to compress/hash the patterns, sort the result hash values, and download them to the HBT nodes. It is a very simple process that fits the requirement for a dynamic update.

3.7.2 Parallel Execution

The HBT can read a single input word per clock cycle, only. When a host interface allows to transfer N bytes of data per clock cycle, a single HBT needs N clock cycles to inspect the input data with a one-byte resolution. To avoid waiting states, N HBT processors should be implemented in parallel (Figure 3.5).

3.7.3 Hardware Implementation of the Tree

It can be easily noticed, from the pipeline execution concept described earlier, that at a certain processing step only one node is busy at each level of the tree. As hardware architecture is the goal, it can hardly be accepted, as it caused unnecessary resource underutilization. The hardware synthesis technique, known as register ports sharing, can be used to optimize the tree. All registers belonging to the same level of the tree are gathered into a single memory block. Then, the corresponding execution hardware can be reduced to a single unit. The register sharing allows it to share operation units between nodes that belong to a single level of the tree. Additionally

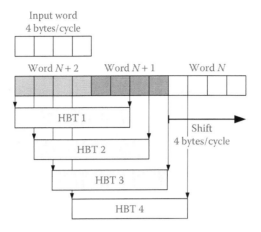

FIGURE 3.5   Example of parallel execution of 4 HBTs.

this approach leads to the opportunity of utilizing the FPGA's memory resources to lessen the amount required by the Flip-Flops for the algorithm implementation. The idea is presented in Figure 3.6.

If a match occurs at a certain level, the hit information is placed in a HIT field. It consists of a level number *l* and the appropriate node address *addr* at that level. From this information, the pattern index can be easily derived. The process of the pattern index calculations is performed in the hardware. The index can be used directly to retrieve adjacent pattern for further inspection.

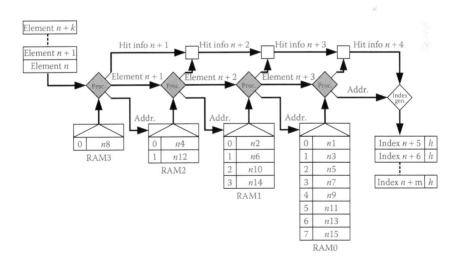

FIGURE 3.6   Pipeline architecture of HBT hardware.

3.8 RESULTS

To measure performance of the presented solution, respective parts of the algorithm were implemented in hardware or software. The goal was to compare throughput of hardware and software subsystems. Stage 1 of the algorithm was implemented in the FPGA, which allowed the evaluation of necessary logic resources. Stages 2–5 were coded in ANSI C. Additionally, to let HW–SW implementation be useful, virus database preprocessing is necessary (as described in Section 3.5.2). Both ClamAV preprocessing and the results of implementations will be presented next.

3.8.1 ClamAV Preprocessing

The proposed algorithm was applied to ClamAV virus database. The tested database version consisted of 94,936 body-based viruses. Results of a test are summarized in Table 3.1. The static substrings (traits) were searched for different length $t = \{16, 32, 48, 64, 80, 96\}$ bits. The column (b) of Table 3.1 presents the number of viruses that do not contain a unique trait in its body. This number depends on t.

TABLE 3.1   Results of Trait Extraction and Hashing for 94,936 Body-Based Viruses

Traits Search		Hashing		Number of Unique Hashes 94,936-(b)-(d) (virs.)
Trait Length (bits)	No Unique Trait (virs.)	Hash Length (bits)	No Unique Hash (virs.)	
(a)	(b)	(c)	(d)	(e)
16	94,936	–	0	0
32	6,254	16	40,644	48,038
48	4,398	16	41,958	48,580
		32	288	90,250
64	4,056	16	41,902	48,978
		32	199	90,681
		48	153	90,727
80	3,995	16	42,110	48,831
		32	182	90,759
		48	133	90,808
		64	123	90,818
96	4,129	16	41,943	48,864
		32	162	90,645
		48	138	90,660
		64	123	90,684
		80	121	90,686

A lack of the ability to find a unique trait for a virus usually occurs when all static substrings of the virus body are similar to the substrings of other viruses. In other situations, the signature is too short when compared to the trait length and no static substring can be selected at all. It can be noted that the smallest number of rejected viruses occurs for $t = 80$ bits. For shorter traits, the number of viruses that have no unique traits increases. The reason is that for shorter traits, the number of possible trait variations decreases. Simultaneously, for longer traits ($t = 96$), the number of rejected viruses also increases. Here, the problem is that it is difficult to extract a static substring that is too long from virus bodies that are segmented with wildcards *.

The table also gives the results of traits hashing. For different traits' length for the hash length $h = 16, 32, 48, 64,$ and 80 bits, where $h < t$, the number of unique hashes is presented. To gain the best system performance, we are interested in such a combination of t and h that maximizes the number of unique hashes that identify with the corresponding viruses' bodies. The unique hash means that the binary data can be checked against a single virus body only. If a single hash is common for two or more viruses, the system has to check all possible viruses that fit within a calculated hash.

The presented example of preprocessing of ClamAV database shows that it is possible to reduce the number of bits necessary to represent viruses in a unique way. For example, if someone devotes 16 bits to represent a single virus then for $t = 16$ no single unique trait exists. Conversely, when a two-phase algorithm is applied then it is possible to achieve 48,976 unique features (hashes) for trait length $t = 64$ bits and $h = 16$ bits. A number of 90,681 unique values can be gained for $t = 64$ and $h = 32$ bits.

3.8.2 Implementation Results

The HBT custom processor was implemented in Xilinx Kintex-7 FPGA technology. The architecture was described in VHDL. The code was synthesized and implemented using Xilinx Design Suite 14.2 tools. The implemented tree consisted of $L = 17$ levels. This number of levels allows it to fill a structure with up to $n = 2^{17} - 1 = 131,071$ patterns. The bit-width of hashes was $h = 32$. The chosen IO interface was a 32-bit AXI4-Stream that is a part of AMBA 4.0, an embedded system bus architecture.

The implementation results of design were as follows:

- Device: xc7k325tfbg676-3

- Logic utilization:

- Number of Slice Flip Flops: 1,677 out of 407,600 (<1%)

- Number of 4 input LUTs: 1,701 out of 203,800 (<1%)

- Number of RAMB36E1s/FIFO36E1s: 127 out of 445 (28%)

- Clock frequency: 133 MHz

- Power consumption: 1.3 W

It can be noticed that the HBT architecture requires a few general logic resources. The utilization of registers and LUTs is below 1%. The main requirement of the HBT is the FPGA's dedicated memory blocks, that is, BRAMs. The architecture can process a single pattern at each clock cycle. For a given clock frequency, an HBT throughput is $r = 1.064$ Gbps.

Now, for the given HBT metrics we will calculate an expected *hit* rate. The rate determines the number of patterns that have to be verified by the software part of the algorithm. We assume a number of viruses $n = 100,000$, hash size $h = 32$, Bloom memory size $M = 2h$, and $k = 1$. According to Equation 3.1 $p_{err} = 2.32 * 10^{-5}$. Because our architecture can process one new byte at each clock cycle (the single was HBT implemented), the *hit* rate is $r_{hit} = (p_{err} * r) \div 8 = 3,085$ hits per second.

3.8.3 Software Stages Performance

Stages 2–5 of the algorithm were coded in the ANSI C language. The program was compiled and run on a system equipped with an ARM Cortex-A9 CPU core. The PetaLinux 12.9 [28] operating system was used to manage system resources. To achieve a moderate size of the index table, the index hash size was set to $x = 12$ bits. The virus database was stored in an external DDR3 memory. The goal was to determine the performance of the software execution. We had to determine if it fits with the hardware processor's output hit rate. It should be noted that the performance of the software algorithm is not deterministic. It varies with the complexity of the viruses' bodies. In this case, an average performance must be established. We measured the runtime necessary for the positive verification of all virus patterns from the database. It should be noted that positive verification is more computationally exhaustive than negative verification because in the case of negative virus verification not all algorithm's stages must be executed.

The software system positively verified 90,681 viruses in 3.5 seconds. The aforementioned means that the CPU is able to verify over $r_{ver} = 25,000$ patterns per second. This corresponds with the achieved r_{hit} as verification rate is higher than hit rate ($r_{hit} < r_{ver}$).

The power consumption of the Cortex-A9 CPU is approximately 1.5 W.

3.9 CONCLUSIONS

The suggested platform for the execution of the 5-stage algorithm is the Zynq-7000 SoC. The system employs the ARM Cortex-A9 CPU and HBT processor. The CPU performance is well balanced with the hardware accelerator throughput. The CPU execution time of virus verification has enough headroom to cope with abnormal bursts of *hit* occurrences. In such a dedicated system, there is no reason to use a more powerful CPU.

The total energy consumption of 3.0 W is very modest. The solution is well suited for distributed clusters of energy-efficient servers such as Moonshot. This kind of processing of the large number of network queries has become very popular recently.

When compared to the performance of the FPGA systems presented in literature where throughput varies from 1 to 11.3 Gbps, the speed of a single HBT is modest. However, the Zynq-7000 family offers a xc7z045 device with 545 RAMB36E1s/FIFO36E1s components. This amount would allow to implement 4 HBTs, which could work in parallel. It is suitable with a standard 32-bit width of AXI-Stream interface. As a result, the theoretical throughput of 4.526 Gbps is possible. The assumed CPU can easily handle the output of the four HBT accelerators working in parallel. On the other hand, when a system network connection is cabled with copper, using the most popular 1000BASE-T standard, then 1 Gbps transfer limit occurs. This suggests that there is no need to strengthen the system performance by adding additional HBTs.

Thanks to the HW–SW approach, our solution outperforms other custom architectures in terms of a volume of stored viruses. The approaches presented in Section 3.3 suffer a small number of viruses that can be handled. Even the total number of coded characters was an issue in other works. Here, we had a limit of $n = 2^{17}$ virus bodies, and the total number of characters is not an issue as the virus database is stored in DDR3 RAM.

ACKNOWLEDGMENTS

This work was supported by the National Science Centre (NCN) [grant number DEC-2011/01/B/ST6/03024] and the National Centre for Research and Development (NCBiR) [grant number SP/I/1/77065/10].

REFERENCES

1. Snort. Snort network intrusion detection system, http://www.snort.org/ [Accessed February 1, 2013], 2013.
2. Bro. Bro intrusion detection system, http://www.bro-ids.org/ [Accessed February 1, 2013], 2013.
3. ClearFoundation. L7-filter, http://l7-filter.clearfoundation.com/ [Accessed: February 1, 2013], 2013.
4. ClamAV. Clam antivirus signature database, http://www.clamav.net/ [Accessed February 1, 2013], 2013.
5. Hewlett-Packard. HP project moonshot, *4AA3-9839ENW.pdf* [Accessed February 1, 2013], 2013.
6. E. Jamro, P. Russek, A. Dabrowska-Boruch, M. Wielgosz, and K. Wiatr. The implementation of the customized, parallel architecture for a fast word-match program. *Comput. Syst. Sci. Eng.*, 26(4):285–292, July 2011.
7. P. Russek and K. Wiatr. The enhancement of a computer system for sorting capabilities using FPGA custom architecture. *Comput. Inform.*, 32(4):859–876, 2013.
8. Xilinx. Zynq-7000 AP SoC Overview, http://www.xilinx.com/ [Accessed February 1, 2013], 2013.
9. C. R. Clark, C. D. Ulmer, and D. E. Schimmel. An FPGA-based network intrusion detection system with on-chip network interfaces. *Int. J. Electr.*, 93(6):403–420, 2006.
10. S. Kumar, S. Dharmapurikar, F. Yu, P. Crowley, and J. Turner. Algorithms to accelerate multiple regular expressions matching for deep packet inspection. *SIGCOMM Comput. Commun. Rev.*, 36(4):339–350, August 2006.
11. R. Sidhu and V. K. Prasanna. Fast regular expression matching using FPGAs. In *Proceedings of the 9th Annual IEEE Symposium on Field-Programmable Custom Computing Machines*, 2001.
12. B. L. Hutchings, R. Franklin, and D. Carver. Assisting network intrusion detection with reconfigurable hardware. In *Proceedings of the 10th Annual IEEE Symposium on Field-Programmable Custom Computing Machines*, page 111, Washington, DC, 2002. IEEE Computer Society.
13. C.-H. Lin, C.-T. Huang, C.-P. Jiang, and S.-C. Chang. Optimization of regular expression pattern matching circuits on FPGA. In *Proceedings of the Conference on Design, Automation and Test in Europe: Designers' Forum*, pages 12–17, Leuven, Belgium, 2006. European Design and Automation Association.
14. I. Sourdis, J. Bispo, J. M. Cardoso, and S. Vassiliadis. Regular expression matching in reconfigurable hardware. *J. Signal Process. Syst.*, 51(1):99–121, April 2008.
15. Y.-K. Chang, C.-R. Chang, and C.-C. Su. The cost effective pre-processing based nfa pattern matching architecture for nids. In *Proceedings of the 2010 24th IEEE International Conference on Advanced Information Networking and Applications*, pages 385–391, Washington, DC, 2010. IEEE Computer Society.

16. Y.-H. E. Yang, W. Jiang, and V. K. Prasanna. Compact architecture for high-throughput regular expression matching on FPGA. In *Proceedings of the 4th ACM/IEEE Symposium on Architectures for Networking and Communications Systems*, pages 30–39, New York, 2008. ACM.

17. X. Wanga, Z. Wang, D. Chen, F. Jiang, and L. Xu. Comparing different approaches to model error modeling in robust identification. *J. Inform. Comput. Sci.*, 9(6):1741–1748, 2012.

18. J. Bispo and J. M. P. Cardoso. Synthesis of regular expressions for FPGAs. *Int. J. Electr.*, 95(7):685–704, 2008.

19. J. M. Bande Serrano and J. H. Palancar. String alignment pre-detection using unique subsequences for FPGA-based network intrusion detection. *Comput. Commun.*, 35(6):720–728, March 2012.

20. M. Bando, N. Sertac Artan, and H. Jonathan Chao. Scalable lookahead regular expression detection system for deep packet inspection. *IEEE/ACM Trans. Netw.*, 20(3):699–714, June 2012.

21. B. H. Bloom. Space/time tradeoffs in hash coding with allowable errors. *Commun. ACM*, 13:422–426, 1970.

22. S. Dharmapurikar, P. Krishnamurthy, T. S. Sproull, and J. W. Lockwood. Deep packet inspection using parallel bloom filters. *IEEE Micro.*, 24(1):52–61, January 2004.

23. J. W. Lockwood, J. Moscola, M. Kulig, D. Reddick, and T. Brooks. Internet worm and virus protection in dynamically reconfigurable hardware. In *Proceedings of the Conference on Military and Aerospace Programmable Logic Device (MAPLD)*, page 10, 2003.

24. S. Dharmapurikar and J. Lockwood. Fast and scalable pattern matching for network intrusion detection systems. *IEEE J. Sel. Areas Commun.*, 24(10):1781–1792, 2006.

25. J. Ho and G. Lemieux. Perg: A scalable FPGA-based pattern-matching engine with consolidated bloomier filters. In *International Conference on ICECE Technology, 2008. FPT 2008,* pages 73–80, December 2008.

26. J. Tsung Lin Ho and G. G. F. Lemieux. Perg-rx: A hardware pattern-matching engine supporting limited regular expressions. In *Proceedings of the ACM/SIGDA International Symposium on Field Programmable Gate Arrays*, pages 257–260, 2009.

27. T. H. Cormen, C. Stein, R. L. Rivest, and C. E. Leiserson. *Introduction to Algorithms*. McGraw-Hill, 2001.

28. PetaLogix. Petalinux, http://www.petalogix.com/ [Accessed February 1, 2013], 2013.

CHAPTER 4

Case Study of Genome Sequencing on an FPGA

Survey and a New Perspective

Chao Wang, Peng Chen, Xi Li, Xiang Ma,
Qi Yu, Xuehai Zhou, and Nadia Nedjah

CONTENTS

4.1 INTRODUCTION

With the insight that biologically significant polymers such as the proteins and the DNA can be abstracted into character strings (sequences), the sequencing technology provides unprecedented opportunities for the life science research. In recent years, next-generation sequencing (NGS) technology has brought great innovations to many applications in the bioinformatics, such as the genome resequencing for single nucleotide polymorphisms (SNPs) detection, the whole-genome expression profiling, the small ribonucleic acid (RNA) discovery, and the development of personalized medicine [1]. Inevitable to all these applications is the procedure of sequence alignment [2], whereby newly sequenced genome fragments are compared to the known subject database on a large scale. The job has been the bottleneck of many bioinformatics researches for a long time not only for the large amount of the sequence reads (always at the magnitude of billions) but also for the enormous long length of the database reference (e.g., three billion base pairs for the human genome). Furthermore, the tolerance for the genome variations and gaps (the addition and deletion of the genomes) in addition increases the difficulty of the alignment. Inspiringly, it has been reported by survey [3] that the state-of-the-art aligners are already fast enough to deal with the ultra amount of short reads (<400 bp) in normal time. In other words, the aligners could keep pace with the generation speed of the sequencing machines. The sequence alignment is not the bottleneck in the genome analysis anymore. However, the researchers are facing new challenges [3] for the NGS long read mapping. With the advent of the new sequencing technologies, which makes it possible to generate much longer reads than before, many current short read aligners will be not applicable in a foreseeable future. In the meantime, current long read mapping tools are not powerful enough to handle the reads efficiently. Furthermore, both the read length and the read amount are constantly increasing with the improvement of the sequencing technology. According to the statistics [4,5], the length of reads produced by the Roche/454 machines has increased from about 250 bp in 2007 to about 500 bp in 2009, and the length of reads generated by the Illumina machines has increased from about 30–50 bp in 2007 to about 100 bp in 2009. The Pacific Biosciences claims that it can generate reads as long as about 1,000 bp in 2009. Besides, it is very important and challenging to align thousands or even millions of base pairs of reads to the genome databases.

Figure 4.1 illustrates the summary of the alignment algorithms in the past decades. It can be derived that more than 10 new approaches have come into market and academics during the past few years.

In respect to the critical situation that long sequence reads will dominate the sequencing field in the near future, while current aligners could still not handle the tremendous amount of reads in the reasonable time, we propose a novel PC-FPGA hybrid architecture in this chapter to improve the performance of the long read mapping.

To address the above problems, in this chapter, we present a heterogeneous cloud framework with MapReduce and multiple hardware execution engines on FPGA to accelerate the genome sequencing problems. The contribution of this chapter is claimed as follows:

1. We propose a novel architecture to accelerate read mapping processing thread when facing large amount of requests with FPGA-based MapReduce.

2. A Distributed MapReduce framework is presented to dispatch the task into multiple FPGA accelerators. The Map and Reduce process is based on RMAP sequencing algorithm. And inside each FPGA-based accelerator, we implement BWA-SW algorithm kernel in hardware that is connected to a local microprocessor to speed up the local alignment process.

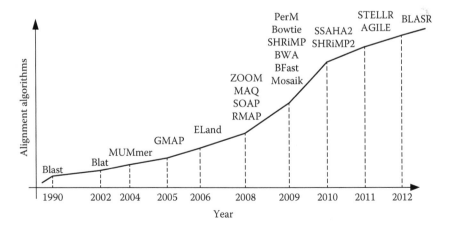

FIGURE 4.1   Summary of the alignment algorithms in past decades.

3. We conduct a theoretical analysis on the MapReduce framework, and for each FPGA chip, we construct a hardware prototype subsystem. The experimental results with hardware speedup, sensitivity, quality, error rate, and hardware utilization are presented and analyzed.

The structure of this chapter is organized as follows. We present the motivation and the related state-of-the-art studies in Section 4.2; then Section 4.3 illustrates the architecture and FPGA-based MapReduce programming framework. We also detail the execution flow of short read mapping in Section 4.3. Thereafter, we explore the hot spots of the genome sequencing application in Section 4.4. A theoretical study for both timing complexity and speedup analysis is also described in Section 4.4. In Section 4.5, we present the hardware prototype and detail the experimental results on the real FPGA hardware platform. Finally, Section 4.6 concludes the chapter and introduces some future works.

4.2 MOTIVATIONS AND ALIGNERS

There are quite a lot of successful short read mapping software tools that address the problem of processing the enormous amount of data produced by the NGS machines. In this section, we first present the motivation, and then analyze the state-of-the-art related work of short read mapping algorithms and sequence aligners.

4.2.1 Motivation

Reads are habitually called short reads if their lengths are shorter than 200 base pairs and called long reads for the opposite [3]. As the limitation of the sequencing technology in history, reads produced by the mainstream sequencers are mostly shorter than 200 base pairs, namely, the short reads. As a result, many of the state-of-the-art aligners present are naturally optimized aimed at the short read alignment. Unfortunately, most of these aligners will not be applicable when the lengths of the reads become long.

For general short read aligners, long length reads would greatly hurt their efficiency and throughput. Some short read aligners just functionally cannot operate properly when the reads lengths exceed certain base pairs. Before explaining more detail why there is a big performance loss when trying to align long reads with the short read aligners, we would give a brief introduction about the mainstream aligners first. The sequence mapping tools can be divided into two major categories based on their index constructing paradigm [3,6].

The first category algorithms construct auxiliary data structures for the genome reference or the sequence reads based on hash tables. The idea of the hash table-based aligners can be traced back to the BLAST [7]. This category of algorithms essentially follows the seed-and-extend paradigm. A fact can be leveraged that the sequence reads are often partially matched in the database. In other words, it is very likely that there must be some short subsequences of a read, which exactly match the genome reference. Thus, by filtering places with partially matched regions, the search scope for the whole-read matches can be greatly reduced. Generally, subsequences are chosen from the reference with certain strategy at first to act as the seeds. A hash table is organized with these seeds being the key and their positions in the reference as the hash value. Then subsequences are fetched from each read with the same strategy and a scan against the entire reference is performed for these subsequences for the exact match by looking up the hash table. The procedure will yield a set of candidate alignment locations (CALs) where partial regions of the reads have matched in the reference and very possible to be the true alignment places. Next, the reads will be extended at each of the CALs and a similarity score between the extended strings from the read and the reference will be carried out with a local alignment algorithm. Finally, only the locations with enough high similarity scores are regarded as the alignment places. Aligners, such as the Eland [8], the SOMP [9], the MAQ [10], and the BFAST [11], are all hash-based mapping tools. Figure 4.2 gives an example to demonstrate the difficulty for matching the long length reads with short read aligners of this category. The strings with the rectangular outline in Figure 4.2a, b are the references while the sequences below them are the reads to align. The references as well as the total data amount ($m = 90$ bp) of the reads in Figure 4.2a, b are the same, yet there are more number of reads with shorter length in Figure 4.2a and less number of reads with longer length in Figure 4.2b. The time used by the alignments will be different for Figure 4.2a, b even though they have the same reference and data amount. Assume that the seeds' length k equals 3 ($k = 3$). Three seeds (seeds quantity $q = n - k + 1$, where n represents the length of the reads and k is the length of the seeds) can be fetched from every read in Figure 4.2a while 13 seeds can be fetched from every seed in Figure 4.2b. Thus it needs totally $(m/n) \cdot (n - k + 1) = 18 \times 3 = 54$ (m

TCTTTGGGAGGAGTGTGGTTTCCTAGACACGGTCACACAATCACTCA

```
                ATTGT         TCCTA       ACGGT    CACAA
    TCTTT     AGGAT  GTGGT           GGCAC    TCACA
    CTTTG      GGATT     GTTTC      GCACG          CACAA
       TGGGA       TTGTG        CTAGG    CGGTC
```

(a)

TCTTTGGGAGGAGTGTGGTTTCCTAGACACGGTCACACAATCACTCA

```
    TCTTTGGGAGGATTG              CTAGGCACGGTCACA
       GGAGGATTGTGGTTT       TAGGCACGGTCACAC
            TTGTGGTTTCCTAG              TCACACAATCACTCA
```

(b)

FIGURE 4.2 Difficulty for matching the long length reads with short read aligners: (a) align 5bp reads versus a 47bp reference ($m = 90$, $n = 5$) and (b) align 15bp reads versus a 47bp reference ($m = 90$, $n = 15$).

indicates the total reads base pairs; $m = 90$ in the example) times of scans for Figure 4.2a and $6 \times 13 = 78$ times of scans for Figure 4.2b. Moreover, the seed extension strings in Figure 4.2b are about three times longer than that in Figure 4.2a, which will further increase the time cost in the local alignment phase. The difference can become very apparent for long reads, which are always hundreds of times longer than the short reads.

The second category sequence alignment algorithms employ the prefix tree or the suffix tree to find the local match locations. This kind of algorithms is always memory efficient as they store the index of the reference or the reads using the Burrows–Wheeler Transform algorithm (an excellent block sorting data compression algorithm). Sequence read mapping tools, such as the MUMmer [12], the Bowtie [13], and the BWA [14], are aligners belonging to this category. These algorithms traverse the sequence reads in the prefix or suffix tree of the reference to look for the partially matched places with a tolerance of a limited number of differences. Then, various approaches are adopted to filter out the true alignment. As for such category of the sequence alignment algorithms, the procedure of traversing the prefix or suffix tree of the reference will cost more time with the increase of the reads lengths. As a result, the overall performance of the aligners will be reduced significantly when the read lengths become tens or hundreds of times longer.

Learned from the discussion above, it is aware that short read aligners will not be efficient when the genome reads become longer. Facing the tread that the reads lengths keep constantly increasing, it is really a critical work to enhance the performance and the efficiency of the long read mapping.

4.2.2 Current Short Read Mapping Algorithms

In this section, we explore the state-of-the-art short read mapping solutions, in particular two main algorithmic categories, as is inherited from Olson et al. [6]. The first category of solutions is based upon a block sorting data compression algorithm called the *Burrows–Wheeler Transform* [15]. These solutions use the FMindex [16] to efficiently store information required to traverse a suffix tree for a reference sequence. These solutions can quickly find a set of matching locations in a reference genome for short reads that match the reference genome with a very limited number of differences. However, the running time of this class of algorithm is exponential with respect to the allowed number of differences; therefore, BWT-based algorithms tend to be less sensitive than other creditable solutions. Bowtie [17] and BWA [14] are example programs based on this algorithmic approach.

The second category of solutions leverages the fact that individual genomes differ only slightly, meaning it is likely that some shorter subsequences of a short read will exactly match the reference genome. This technique is called seed-and-extend, and these shorter subsequences are called seeds. In fact, if a we align an m bp read with at most k differences, there must exist one exact alignment of $m/(k + 1)$ consecutive bases [18]. An index of the reference genome is compiled first, which maps every seed that occurs in the reference genome to the locations where they occur. To align a short read, all the seeds in the read are looked up in the index, which yields a set of CALs. The short read is then scored against the reference at each of these CALs using the Smith–Waterman [19] string-matching algorithm. The location with the highest score is chosen as the alignment location for a short read. For example, BLAST [11] uses a hash table of all fixed length k-mers in the reference to find seeds, and a banded version of the Smith–Waterman algorithm to compute high scoring gapped alignments. RMAP [20] uses a hash table of nonoverlapping k-mers of length $m/(k + 1)$ in the reads to find seeds.

Based on the string match problem abstractions, plenty of traditional approaches those tackle high-speed similarity analysis and indexing exploration can be adopted [21–25]. Baker and Prasanna [26] presented a hardware implementation of the Knuth–Morris–Pratt (KMP) algorithm. Since the KMP algorithm is designed to match the input stream against a single string, one matching unit is required per string, and the hardware system is composed of a linear array of matching units. The methods presented in Baker and Prasanna [27] and Dimopoulos et al. [28] are based on pre-decoded characters with hardwired logic circuits. The system is optimized with respect to the given set of signatures and the characteristics of the FPGA devices. The FPGA is reconfigured when there are changes to the signature set. However, the long latency required in offline generation of the optimized hardwired circuits is considered as a major disadvantage in a network intrusion detection system that demands fast responses to hostile conditions. For the sake of FPGA accelerators, many of the proposed hardware solutions are based on the well-known Aho–Corasick (AC) algorithm [29], where the system is modeled as a deterministic finite automaton (DFA). The AC algorithm solves the string matching problem in time linearly proportional to the length of the input stream. However, the memory requirement is not feasible in a straightforward hardware implementation. In particular, Pao et al. [30] present a pipelined processing approach to the implementation of AC algorithm. Other published AC-based methods are heuristic-based, such as bit-map encoding and path compression [31], bit-slice implementation [32], categorizing alphabets based on frequency count [28], and signature set partitioning [33]. However, these approaches are not appropriate to implement in hardware due to the complexity and timing overheads.

4.2.3 State-of-the-Art Sequence Aligners

Exploring the secret of the genomes of the beings is a promising job with seemingly limitless possibilities to get to know the truth of the life. Sequence alignment, as one of the fundamental biotechnologies in life science, has drawn a lot of attention from many groups. Table 4.1 lists some of the well-known sequence aligners in history as well as their partial features.

Entries in Table 4.1 are listed as far as possible in chronological order. The second column gives the references of the aligners. Indels caused by the additions or deletions of the genomes can often be found in the data

TABLE 4.1   State-of-the-Art Sequence Aligners

Aligners	Ref	Gapped	PE	Length	Technique
BLAST	[7]	Yes	No	Short	HSP[a]
BLAT	[34]	Yes	No	Both	k-mers
MUMmer	[12]	Yes	No	Both	Suffix tree
GMAP	[1]	Yes	No	Both	Sandwich DP[b]
Eland	[8]	No	No	Short	k-mers
RMAP	[20]	No	Yes	Short	Quality scores
SOAP	[9]	Yes	Yes	Short	Hash table
MAQ	[10]	Yes	Yes	Short	Quality scores
ZOOM	[2]	Yes	Yes	Short	Spaced Seed
SHRiMP	[35]	Yes	Yes	Short	S–W[c]
BFAST	[11]	Yes	Yes	Short	S–W
PerM	[36]	No	Yes	Short	Spaced Seed
Bowtie	[13]	No	Yes	Short	BWT[d]
BWA	[14]	Yes	Yes	Short	BWT
Mosaik	[37]	Yes	Yes	Both	S–W
BWA-SW	[38]	Yes	Yes	Long	DAWG[e], S–W
SSAHA2	[39]	Yes	Yes	Both	S–W
AGILE	[4]	Yes	No	Long	q-gram
STELLR	[40]	Yes	No	Long	SWIFT, ε-core

[a] HSP: High-scoring segment pair.
[b] DP: Dynamic programming.
[c] S–W: Smith–Waterman algorithm.
[d] BWT: Burrows–Wheeler transform.
[e] DAWG: Directed acyclic word graph.

produced by sequencers. To tolerate the indels in the reads, many aligners integrate the ability to deal with gapped data (listed in the third column of Table 4.1). Besides, some of the sequencers can generate reads from both ends of the genome fragments. Such reads (called pair-end reads) are always in pairs and often have settled relative orientation and approximate distance between each other. With respect to such a situation, some aligners provide the support for the pair-end (PE) reads (the fourth column in Table 4.1). The fifth column in Table 4.1 indicates whether the aligner is efficient for the short reads and the long reads. The last column gives a brief note of some technical points used by each aligner.

4.3 ACCELERATING APPROACHES

Along with the novel short read mapping algorithms, multiple attempts have also been conducted to accelerate short read mapping in diverse high performance computing techniques. Graphics processing units (GPUs), MapReduce, and FPGA-based hardware are the most widely used acceleration engines. Table 4.1 listed most of the state-of-the-art literatures for type, reference, performance metric, utilized algorithm, and special features. In particular, the devoted researches can be divided into following categories.

4.3.1 GPU-Based Accelerations

GPUs have recently been used as a mature approach for several bioinformatics applications, especially for sequence alignment, one of the most significant research areas. For example, Manavski and Valle [41] presented a similarity using Smith–Waterman algorithm in GPU accelerators, using compute unified device architecture (CUDA) programming engines. MUMmerGPU [42] is an open-source parallel pairwise local sequence alignment program that runs on commodity GPUs in common workstations. Based on this research, Trapnell and Schatz [43] used a stackless depth-first-search print kernel with massive GPU data layout configurations to improve register footprint and conclude higher occupancy. Liu et al. [44] made new contributions to Smith–Waterman protein database searches using CUDA. A parallel Smith–Waterman algorithm has been proposed to further optimize the performance based on the single-instruction-multiple-thread (SIMT) abstraction.

4.3.2 FPGA-Based Accelerations

FPGA devices play an important role in the acceleration of sequence alignment. As early as the beginning of this century, an FPGA-based platform called Tera-BLAST [45] is developed by TimeLogic Inc. to enhance the performance of the BLAST mapping tool. A scalable accelerator for comparing the protein sequences is brought forward in Faes et al. [46], which implements the Smith–Waterman–Gotoh algorithm and is able to align two sequences of length 1024 base pairs. The Mercury BLASTP [47], which is implemented on a workstation with two Xilinx Virtex-II 6000 FPGAs, could obtain a 11–15× speedup over the BLASTP software while delivering close to 99% identical result. A family of algorithms came up in Van Court and Herbordt [48], which can generate 256 different accelerators for approximate string matching. Besides, there are more solutions [6,49–52] in recent years to accelerate the read mapping. However, most of these solutions are targeting the short read mapping and are applicable for reads with lengths of several tens or under several hundreds. Honestly speaking, there are some aligners [46,50,51], though originally designed for the short read mapping, that could align several hundreds or even thousands of base pairs of reads. Unfortunately, the software aligners they are accelerating are not so efficient compared to the state-of-the-art ones. Reports show that some more recently developed aligners have a huge improvement in the aspect of performance [3]. For example, the BWA-SW aligner [38], one of the new generation long read aligners, hurts the traditional long read aligners such as SSAHA2 and BLAT in terms of both speed and accuracy. As a newly developed long read aligner, BWA-SW aligner is claimed to be as accurate as SSAHA2, more accurate than BLAT, and several to tens of times faster than both [38]. The performance improvement of the new aligners could cover the speedup of many hardware acceleration solutions. Overall, both the long lengths of the reads and the huge performance improvement of the new generation aligners motivate the study of the new hardware architecture for accelerating the long read mapping. We have a previous work [53], which introduces the hardware acceleration for the long read mapping. Wang et al. [54] tentatively used MapReduce framework to accelerate the long read mapping problems.

Nevertheless, numerous attempts to accelerate short read mapping on FPGAs tried to use a brute-force approach to compare short sequences in parallel to a reference genome. For example, the work in Knodel et al. [55] and Fernandez et al. [56] streamed the reference genome through a system doing exact matching of the short reads. Knodel et al. [55] demonstrated a greater sensitivity to genetic variations in the short reads than Bowtie and MAQ, but the mapping speed was approximately the same as that of Bowtie. Also, this system demonstrated mapping short reads to only chromosome 1 of the human genome. Fernandez et al. [56] demonstrated between 1.6× and 4× speedup versus RMAP [20] for reads between 0 and 3 differences. This implementation was for the full human genome.

Some recent works conducing DNA short read sequencing problem using FPGA-based reconfigurable computing acceleration engines are proposed [6,49,50,57–60]. Of these approaches, Preusser et al. [49] proposed a systolic custom computation on FPGA to implement the read mapping on a massive parallel architecture. This literature enables the implementation of thousands of parallel search engines on a single FPGA device. Tang et al. [57] proposed a CPU-FPGA heterogeneous architecture for accelerating a short reads mapping algorithm, which was built upon the concept of hash index with several optimizations that reorder hash table accesses and compress empty hash buckets. Mahram and Herbordt [50] applied prefiltering of the kind commonly used in BLAST to perform the initial all-pairs alignments. Olson et al. [6] proposed a scalable FPGA-based solution to the short read mapping problem in DNA sequencing, which greatly accelerates the task of aligning short length reads to a known reference genome. Not only the first stage of progressive alignment in multiple sequence alignment problem has been investigated; the third stage of progressive alignment on reconfigurable hardware in Lloyd and Snell [58] has also been attractive in the state-of-the-art researches. Chen et al. [59] introduced a hybrid system for short read mapping, utilizing both software and FPGA-based hardware. Zhang et al. [60] presented an implementation of the Smith–Waterman algorithm for both DNA and protein sequences on the platform. The article introduces a multistage processing element design and a pipelined control mechanism with uneven stage latencies to improve the performance of and decrease the on-chip SRAM usage. Chen et al. [61] presented a novel FPGA-based architecture, which could address the problem with a bounded number of PEs to realize any lengths of systolic array. It is mainly based on the idea of the banded Smith–Waterman but with a key difference that it reuses the PEs that are beyond the boundary.

Previous efforts doing short read mapping using FPGAs have achieved at most an order of magnitude improvement compared to software tools. Also, previous solutions are not convincible to produce a system that is well suited to large-scale long read mapping and full genome sequencing. Finally, current literatures only use one unique method on FPGA to accelerate the short read mapping problem; therefore, it could make the most advantage of the parallel computing and hardware acceleration–based techniques.

4.3.3 MapReduce-Based Acceleration Frameworks

Besides the above heterogeneous accelerating engines, MapReduce [62] is an alternative software framework developed and used by Google to support parallel-distributed execution of data-intensive applications. Utilizing MapReduce programming framework, CloudBurst [63] is a new creditable parallel read mapping algorithm optimized for mapping next-generation sequence data to the human genome and other reference genomes, for use in a variety of biological analyses including SNP discovery, genotyping, and personal genomics. It is modeled after the short read mapping program RMAP [20] and reports either all alignments or the unambiguous best alignment for each read with any number of mismatches or differences. This level of sensitivity could be prohibitively time consuming, but CloudBurst uses the open-source Hadoop implementation of MapReduce to parallelize execution using multiple compute nodes, which means it can deal with larger amount of short reads simultaneously. Similarly, Dai et al.'s work [64] is based on the preprocess of the reference genomes and iterative MapReduce jobs for aligning the continuous incoming reads.

To sum up, even the MapReduce framework could disseminate the read mapping to multiple computing machines, but the execution kernel itself is quite inefficient, so the throughput of the entire system is still worth pursuing.

4.4 SHORT READ MAPPING ON HETEROGENEOUS CLOUD FRAMEWORK

As the MapReduce framework has been successfully integrated into FPGA-based researches [65], in this section, we demonstrate the system architecture of a hybrid heterogeneous system utilizing both software and FPGA-based hardware for real-time short reads mapping service. We will model it as a specific-domain search problem that has been studied well for the past decades.

After sequencing DNA, researchers often map the reads to a reference genome to find the locations where each read occurs. The read mapping algorithm reports one or more alignments for each read within a scoring threshold, commonly expressed as the minimal acceptable significance of the alignment, or the maximum acceptable number of differences between the read and the reference genome.

The motivation of combining the MapReduce and the FPGA is to utilize MapReduce framework for the big data aspect and the FPGA for the acceleration part. In particular, growing with the data amount of genome sequencing, it will be essential to use MapReduce to handle the extremely large amount of data. MapReduce framework, in many occasions, has been long proven and demonstrated as an efficient methodology. In this chapter, to benefit from both MapReduce and FPGA, we construct the MapReduce framework on CPU/FPGA hybrid platforms. Normally the process is divided into the following two steps:

1. The first stage is the MapReduce stage. Scientific researchers can send short reads as a stream into a system through the MapReduce server. As soon as the request is received, it will undergo a general Map scheduling stage, which partitions the entire genome sequencing task into many small jobs, and then distributes them to parallel computing nodes.

2. The second stage is the local alignment in FPGA. Multiple hardware acceleration engines are deployed to speed up the genome sequencing analysis procedure.

Our proposed architecture framework is illustrated in Figure 4.3. Generally the system is constructed on a central cluster server, which is responsible for sequence preprocessing, database access, and user interaction, while multiple hardware acceleration engines are deployed to speed up the genome sequencing analysis procedure.

Throughout this chapter, we use Hbase [66] as the genome database, which is a distributed storage system for random, real-time read/write access to our Big Data for very large tables—with billions of rows and millions of columns. Hbase provides big table-like capabilities on top of Hadoop [67] and HDFS [68]. Data stored in Hbase were divided according to the different entries. Each table may have billions of rows and all the rows in the table were ordered by the key field. The biggest difference between Hbase and other SQL databases is that Hbase is column-oriented. Due to the large size of the tables with millions of columns, each column belongs to column family, which is a logic union of columns. Each cell in Hbase stores multiversions of the data. Since one table can contain hundreds or thousands columns, therefore the overall data volume can be millions. This property is very important for our system design. Each

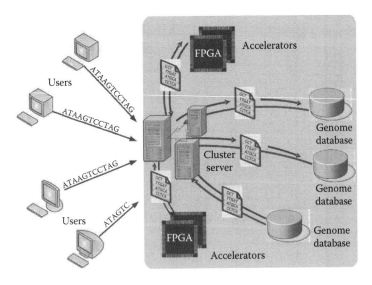

FIGURE 4.3 Architecture framework for hybrid systems.

cell also contains multiversion values that allow us remember the history information.

At start-up, the cluster server is ready to receive multiple user requests at any time. Due to the large amount of short reads (millions at least) in general, it is not practical to start the mapping procedure in one machine, even with the assistance of hardware acceleration engines. We also need to distribute the short reads to different computing machines in parallel, each of which is equipped with FPGA-based accelerators. As a consequence, we extend the MapReduce programming framework with multiple hardware acceleration engines on single FPGA to further reduce the execution time for the big data genome sequencing processing applications.

4.5 THEORETICAL ANALYSIS

In this section, we analyze the timing complexity of the algorithm, and then present the ideal speedup using different hardware architectures.

4.5.1 Timing Complexity Analysis

The terms and parameters used in the equations are defined in Table 4.2. m refers to the minimum short read size, which lies between 100 and 2,000 bp generally; k is the maximum allowed mismatches, which is configured to 4; and s and n represent the seed size and reference genome size, respectively. Due to the large amount of human genomes, the reference genome size could reach to one billion or more. R indicates the number of short reads, which could also be up to one million. Finally, the speedup with multiple hardware accelerators is set to $75 \times c$, while c is the number of accelerators.

Assume that m refers to the minimum short reads size, k stands for the maximum allowed mismatches, s indicates the seed size, n denotes the reference genome size, and R represents the number of short reads, and p refers to the speedup with multiple hardware accelerators. Based on these

TABLE 4.2   Parameters Used to Analyze Speedup

Symbols	Description	Typical
m	Minimum short reads size	100 ~ 2,000 bp
k	Maximum mismatches	4 (<10)
s	Seed size from reads	$m/(k+1)$
n	Reference genome size	1 billion
R	Number of reads	1 million
p	Speedup with multiple HW	$75 \times c$

parameters and the processing flow of the algorithm, the timing complexity is analyzed separately in the following two phases:

1. *Compare phase:* In serial algorithm, each read was divided into m/s seeds, and each seed compares with the reference genome to get index, and the complexity would be $O(n + s)$; therefore, the timing complexity in this stage is $O(mn + m)$.

2. *Seed and extend phase:* After that, we use a variation of the Landau–Vishkin k-difference alignment algorithm that costs $O(km)$ to get the score of each seed. Considering the m/s seeds, the timing complexity in this stage is $O(km^2)$.

So the total time would be $O(m * n + km^2 + m)$. For R reads, the total runtime would be $O(Rmn + Rkm^2 + Rm)$. It is widely applied that the reference genome size n is much bigger than the minimum short read size m; consequently the final timing complexity is $O(Rmn)$.

Based on the timing analysis, we can derive the speedup as we use both hardware accelerators and MapReduce parallel processing framework. Let c be the nodes number of the acceleration engines, and *SPU* refers to the speedup string match hardware implementations; hence we can achieve the speed by multiple hardware acceleration engines under MapReduce framework $p = SPU \times c$. Finally the total time of R reads in our system should be less than $O(Rmn/p)$.

4.5.2 Speedup Analysis

We have evaluated the curve depicting how the speedup of the hybrid system is affected by the number of acceleration engines. Due to the large amount of the genome size n, the final speedup curve is quite linear with the slope coefficient of SPU (in our case, SPU is configured to 75× against software execution, according to the profiling results on preliminary experiments), as is presented in Figure 4.4.

Both the curve for binary tree and ring topologies appear to be extremely close to the ideal linear speedup. The reason is that the topologies take $\log c$ and $c/2$ for translation delays. When c is relatively smaller than the data sets (n and m are quite huge due to human genome data sets), either approach does make noticeable difference. The curve for both binary trees and ring topologies are essentially coincidental with the ideal linear speedup; therefore, the theoretical speedup for

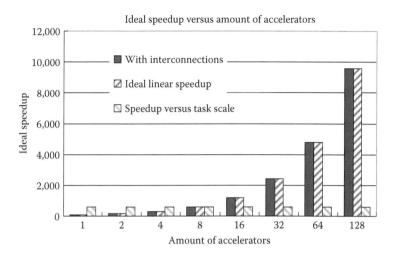

FIGURE 4.4   Theoretical speedup versus amount of accelerators.

64 accelerators can reach to up to 4800× at most. Figure 4.4 also illustrates that the number of accelerators has a linear effect on the speedup when the genome size is much bigger than the short read size. We also measure the curve between speedup and the task scale n. In this case, c is set to 8, and the speedup is fixed to $75 \times 8 = 600$, which is presented in the second bar with left dashes. The results demonstrate that the final speedup is irrelevant to the communication overheads as the application is computing intensive. Due to the liner speedup metrics, it is a favorable to convert parallelly distributed high-level heterogeneous hardware accelerated engines into a sound framework for future biological DNA sequencing solutions.

4.6 CONCLUSIONS

This chapter has proposed a general heterogeneous cloud framework for NGS with multiple hardware accelerators on FPGAs. We utilize an extended MapReduce distribution framework with multiple hardware accelerators on FPGA. By extending a distributed processing technique with hardware accelerators, this approach could bring significant speedup for genome sequencing alignment process. We have presented results from both a theoretical analysis and real hardware platform on Xilinx FPGA development board. The experimental results reveal that each hardware accelerator could achieve up to 2.73× speedup and it only occupy less than 1% of the FPGA chip resources.

Given the promising preliminary results illustrated in this chapter, there exist various directions for future developments. Additional engineering effort needs to be applied to assess the scalability comparison to aforementioned cutting-edge approaches. Also a comparative study between the short read mapping applications in high-performance GPU-based clusters and detailed analysis of the feasibility, cost, and overheads are in progress. Furthermore, from the technical perspective, if different reads contain the same seeds, they could be cached in a fast table to respond with high efficiency.

FUNDING

This work was supported by the National Science Foundation of China [61379040], [61272131], and [61202053]; Jiangsu Provincial Natural Science Foundation [SBK201240198]; Open Project of State Key Laboratory of Computer Architecture; Institute of Computing Technology; Chinese Academy of Sciences [CARCH201407]; and the Strategic Priority Research Program of CAS [XDA06010403].

REFERENCES

1. T. D. Wu and C. K. Watanabe. GMAP: A genomic mapping and alignment program for mRNA and EST sequences. *Bioinformatics*, 2005. **21**(9): pp. 1859–1875.
2. H. Lin, Z. Zhang et al. ZOOM! Zillions of oligos mapped. *Bioinformatics*, 2008. **24**(21): pp. 2431–2437.
3. H. Li and N. Homer. A survey of sequence alignment algorithms for next-generation sequencing. *Bioinformatics*, 2010. **11**(5): pp. 473–483.
4. S. Misra, A. Agrawal et al. Anatomy of a hash-based long read sequence mapping algorithm for next generation DNA sequencing. *Bioinformatics*, 2011. **27**(2): pp. 189–195.
5. B. Vasanth, D. Evelyn et al. Dynamo: A transparent dynamic optimization system. 2000, PLDI. pp. 1–12.
6. C. B. Olson, M. Kim et al. Hardware acceleration of short read mapping. In *Proceedings of the IEEE 20th International Symposium on Field-Programmable Custom Computing Machines*. 2012. pp. 161–168.
7. S. F. Altschul, W. Gish et al. Basic local alignment search tool. *Molecular Biology*, 1990. **215**(3): pp. 403–410.
8. A. CoX. *ELAND: Efficient Local Alignment of Nucleotide Data*. (Unpublished), 2007.
9. R. Li, Y. Li et al. SOAP: Short oligonucleotide alignment program. *Bioinformatics*, 2008. **24**(5): pp. 713–714.

10. R. D. Blumofe, C. F. Joerg et al. Cilk: An efficient multithreaded runtime system. In *Proceedings of the 5th ACM SIGPLAN Symposium on Principles and Practice of Parallel Programming.* 1995, ACM, Santa Barbara, CA, pp. 207–216.
11. N. Homer, B. Merriman et al. BFAST: An alignment tool for large scale genome resequencing. *PLoS ONE,* 2009. **4**(11): p. e7767.
12. S. Kurtz, A. Phillippy et al. Versatile and open software for comparing large genomes. *Genome Biology,* 2004. **5**(2): p. R12.
13. B. Langmead, C. Trapnell et al. Ultrafast and memory-efficient alignment of short DNA sequences to the human genome. *Genome Biology,* 2009. **10**(3): pp. 1–10.
14. H. Li and R. Durbin. Fast and accurate short read alignment with Burrows-Wheeler transform. *Bioinformatics,* 2009. **25**(14): pp. 1754–1760.
15. M. Burrows and D. J. Wheeler. *A Block-Sorting Lossless Data Compression Algorithm.* 1994, Digital Equipment Corporation, Palo Alto, CA. p. 124.
16. P. Ferragina and G. Manzini. Opportunistic data structures with applications. in *Proceedings of the 41st Annual Symposium on Foundations of Computer Science.* 2000, IEEE Computer Society, Redondo Beach, CA, p. 390.
17. B. Langmead, C. Trapnell et al. Ultrafast and memory-efficient alignment of short DNA sequences to the human genome. *Genome Biology,* 2009. **10**(3): p. R25.
18. R. A. Baeza-Yates and C. H. Perleberg. Fast and practical approximate string matching. In *Proceedings of the 3rd Annual Symposium on Combinatorial Pattern Matching.* 1992, Springer-Verlag, Tucson, AZ, pp. 185–192.
19. T. F. Smith and M. S. Waterman. Identification of common molecular subsequences. *Journal of Molecular Biology,* 1981. **147**(1): pp. 195–197.
20. A. D. Smith, Z. Xuan et al. Using quality scores and longer reads improves accuracy of Solexa read mapping. *BMC Bioinformatics,* 2008. **9**(128): pp. 1471–2105.
21. Q. Wang and V. K. Prasanna. Multi-core architecture on FPGA for large dictionary string matching. in *Proceedings of the 2009 17th IEEE Symposium on Field Programmable Custom Computing Machines.* 2009, IEEE Computer Society, Salt Lake City, UT, pp. 96–103.
22. I. Sourdis and D. Pnevmatikatos. Fast, large-scale string match for a 10Gbps FPGA-based network intrusion detection system. In *Proceedings of the 13th International Conference on Field Programmable Logic and Applications.* 2003, Lisbon, Portugal, pp. 880–889.
23. T. Van Court and M. C. Herbordt. Families of FPGA-Based Algorithms for Approximate String Matching. In *Proceedings of the Application-Specific Systems, Architectures and Processors, 15th IEEE International Conference.* 2004, IEEE Computer Society, pp. 354–364.
24. M. Aldwairi, T. Conte et al. Configurable String Matching Hardware for Speeding up Intrusion Detection. *ACM SIGARCH Computer Architecture News,* 2005. **33**(1): pp. 99–107.

25. D. Pao, W. Lin et al. Pipelined architecture for multi-string matching. *IEEE Computer Architecture Letters*, 2008. **7**(2): pp. 33–36.
26. Z. K. Baker and V. K. Prasanna. A computationally efficient engine for flexible intrusion detection. *IEEE Transactions on Very Large Scale Integration (VLSI) Systems*, 2005. **13**(10): pp. 1179–1189.
27. Z. K. Baker and V. K. Prasanna. Automatic Synthesis of Efficient Intrusion Detection Systems on FPGAs. *IEEE Transactions on Dependable and Secure Computing*, 2006. **3**(4): pp. 289–300.
28. V. Dimopoulos, I. Papaefstathiou et al. A memory-efficient reconfigurable Aho-Corasick FSM implementation for intrusion detection systems. In *International Conference on Embedded Computer Systems: Architectures, Modeling and Simulation, IC-SAMOS*. 2007: pp. 186–193.
29. A. V. Aho and M. J. Corasick. Efficient string matching: An aid to bibliographic search. *Communications of the ACM*, 1975. **18**(6): pp. 333–340.
30. D. Pao, W. Lin et al. A memory-efficient pipelined implementation of the aho-corasick string-matching algorithm. *ACM Transactions on Architecture and Code Optimization*, 2010. **7**(2): pp. 1–27.
31. N. Tuck, T. Sherwood et al. Deterministic memory-efficient string matching algorithms for intrusion detection. In *IEEE International Conference on Computer Communications*. 2004, Hong Kong, China, pp. 2628–2639.
32. L. Tan and T. Sherwood. A high throughput string matching architecture for intrusion detection and prevention. In *Proceedings of the 32nd Annual International Symposium on Computer Architecture*. 2005, IEEE Computer Society, Munich, Germany, pp. 112–122.
33. J. van Lunteren. High-performance pattern-matching for intrusion detection. In *IEEE International Conference on Computer Communications*. 2006, Barcelona, Spain, pp. 1–13.
34. W. James Kent. BLAT—The BLAST-like Alignment Tool. *Genome Research*, 2002. **12**(4): pp. 656–664.
35. S. M. Rumble, P. Lacroute et al. SHRiMP: Accurate Mapping of Short Color-space Reads. *PLoS Computational Biology*, 2009. **5**(5): p. e1000386.
36. Y. Chen, T. Souaiaia et al. PerM: Efficient mapping of short sequencing reads with periodic full sensitive spaced seeds. *Bioinformatics*, 2009. **25**(19): pp. 2514–2521.
37. K. Ali, M. Aboelaze et al. Modified hotspot cache architecture: A low energy fast cache for embedded processors. In *Proceedings of the International Conference on Embedded Computer Systems: Architectures, Modeling and Simulation. IC-SAMOS*. 2006, Samos, Greece, pp. 35–42.
38. H. Li and R. Durbin. Fast and accurate long-read alignment with Burrows-Wheeler transform. *Bioinformatics*, 2010. **26**(5): pp. 589–595.
39. Z. Ning, A. J. Cox et al. SSAHA: A fast search method for large DNA databases. *Genome Research*, 2001. **11**(10): pp. 1725–1729.
40. B. Kehr, D. Weese et al. STELLAR: Fast and exact local alignments. *BMC Bioinformatics*, 2011. **12**(Suppl 9): p. S15.

41. S. A. Manavski and G. Valle. A compatible GPU cards as efficient hardware accelerators for Smith-Waterman sequence alignment. *BMC Bioinformatics*, 2008. **9** (Suppl 2): p. S10.
42. M. C. Schatz, C. Trapnell et al. High-throughput sequence alignment using Graphics Processing Units. *BMC Bioinformatics*, 2007. **8**: p. 474.
43. C. Trapnell and M. C. Schatz. Optimizing data intensive GPGPU computations for DNA sequence alignment. *Parallel Computing* 2009. **35**(8–9): pp. 429–440.
44. Y. Liu, B. Schmidt et al. CUDASW++2.0: Enhanced Smith-Waterman protein database search on CUDA-enabled GPUs based on SIMT and virtualized SIMD abstractions. *BMC Research Notes*, 2010. **3**: p. 93.
45. R. Luethy and C. Hoover. Hardware and software systems for accelerating common bioinformatics sequence analysis algorithms. *BioSilico*, 2004. **2**(1): p. 12–17.
46. P. Faes, B. Minnaert et al. Scalable hardware accelerator for comparing DNA and protein sequences. In *Proceedings of the 1st International Conference on Scalable Information Systems*. 2006, ACM, Hong Kong, China, p. 33.
47. B. Harris, A.C. Jacob et al. A banded Smith-Waterman FPGA accelerator for mercury BLASTP. In *Proceedings of the FPL*, 2007, Amsterdam, The Netherlands, pp. 765–769.
48. T. Van Court and M. C. Herbordt. Families of FPGA-based accelerators for approximate string matching. *Microprocess. Microsyst.*, 2007. **31**(2): pp. 135–145.
49. T. B. Preusser, O. Knodel et al. Short-read mapping by a systolic custom FPGA computation. In *Proceedings of the IEEE 20th International Symposium on Field-Programmable Custom Computing Machines*. 2012: pp. 169–176.
50. A. Mahram and M. C. Herbordt. FMSA: FPGA-accelerated clustalW-based multiple sequence alignment through pipelined prefiltering. In *Proceedings of the IEEE 20th International Symposium on Field-Programmable Custom Computing Machines*. 2012, Ontario, Canada, pp. 177–183.
51. X. Guo, H. Wang et al. A systolic array-based FPGA parallel architecture for the BLAST algorithm. *ISRN Bioinformatics*, 2012: p. 11.
52. K. Benkrid, Y. Liu et al. A highly parameterized and efficient FPGA-based skeleton for pairwise biological sequence alignment. *IEEE Transactions on Very Large Scale Integration Systems*, 2009. **17**(4): pp. 561–570.
53. P. Chen, C. Wang et al. Accelerating the Next Generation long read mapping with the FPGA-based system. *IEEE/ACM Transactions on Computational Biology and Bioinformatics*, 2014. **PP**(99): p. 1.
54. C. Wang, X. Li et al. Heterogeneous cloud framework for big data genome sequencing. *IEEE/ACM Transactions on Computational Biology and Bioinformatics*, 2014. **PP**(99): p. 1.
55. O. Knodel, T. B. Preusser et al. Next-generation massively parallel short-read mapping on FPGAs. In *Proceedings of the IEEE International Conference on Application-Specific Systems, Architectures and Processors*. 2011, Santa Monica, CA, pp. 195–201.

56. E. Fernandez, W. Najjar et al. Exploration of short reads genome mapping in hardware. In *Proceedings of the International Conference on Field Programmable Logic and Applications*. 2010: pp. 360–363.

57. W. Tang, W. Wang et al. Accelerating Millions of short reads mapping on a heterogeneous architecture with FPGA accelerator. In *Proceedings of the IEEE 20th International Symposium on Field-Programmable Custom Computing Machines*. 2012, pp. 184–187.

58. S. Lloyd and Q. O. Snell. Accelerated large-scale multiple sequence alignment. *BMC Bioinformatics*, 2011. **12**: p. 466.

59. Y. Chen, B. Schmidt et al. Accelerating short read mapping on an FPGA (abstract only). In *Proceedings of the ACM/SIGDA International Symposium on Field Programmable Gate Arrays*. 2012. Monterey, CA: ACM, p. 265.

60. P. Zhang, G. Tan et al. Implementation of the Smith-Waterman algorithm on a reconfigurable supercomputing platform. In *Proceedings of the 1st International Workshop on High-Performance Reconfigurable Computing Technology and Applications: Held in Conjunction with SC07*. 2007. Reno, Nevada: ACM, pp. 39–48.

61. P. Chen, C. Wang et al. Hardware acceleration for the banded Smith-Waterman algorithm with the cycled systolic array. In *FPT 2013*. 2013, Kyoto, Japan, pp. 480–481.

62. D. Jeffrey and G. Sanjay. MapReduce: Simplified data processing on large clusters. *Communications of the ACM*, 2008. **51**(1): pp. 107–113.

63. M. C. Schatz. CloudBurst: Highly sensitive read mapping with MapReduce. *Bioinformatics*, 2009. **25**(11): pp. 1363–1369.

64. D. Dai, X. Li et al. Cloud based short read mapping service, in *IEEE International Conference on Cluster Computing*. 2012, Beijing, China, pp. 601–604.

65. Y. Shan, B. Wang et al. FPMR: MapReduce framework on FPGA. In *Proceedings of the 18th Annual ACM/SIGDA International Symposium on Field Programmable Gate Arrays*. 2010. Monterey, CA: ACM, pp. 93–102.

66. Apache. *HBase—Apache HBase Home*. 2014, http://hbase.apache.org/.

67. Apache. *Hadoop*. 2014, ttp://hadoop.apache.org/.

68. *HDFS Architecture Guide*. 2014, http://hadoop.apache.org/docs/r1.2.1/hdfs_design.html.

II

Network-on-Chip

Interprocess Communication via Crossbar for Shared Memory Multiprocessor Systems-on-Chip

Luiza de Macedo Mourelle, Nadia Nedjah, and Fábio Gonçalves Pessanha

CONTENTS

5.1 INTRODUCTION

For decades, engineers were trying to improve the processing capability of microprocessors by increasing clock frequency [1]. During the 1990s, the clock frequency of the fasted microprocessors exceeded that of the fastest supercomputers. The next step was to explore parallelism at the instruction level with the concept of pipeline [2]. However, the speedup required by software applications was gradually becoming higher than the speedup provided by these techniques. Besides this, the increase in clock frequency was leading to the increase in power required, to levels not acceptable. The search for smaller devices with high processing capability and with less energy consumption has turned solutions based on only one processor obsolete. This kind of solution has been restricted to low performance applications, for which microcontrollers are the best.

Process-level parallelism is exploited when using multiple processors, running independent programs simultaneously. On the other hand, parallel processing program means a single program that runs on multiple processors simultaneously. In order to reach specific performance requirements, such as throughput, latency, energy consumed, power dissipated, silicon area, design complexity, response time, and scalability, the concept of multicore microprocessors has been explored, where a microprocessor contains multiple processors or cores in a single integrated circuit. Multiprocessor system-on-chip (MPSoC) offers several processors implemented in only one chip to provide the most of parallelism possible. On the other hand, since these processors need to communicate, in order to exchange data required for the parallel execution of the application, a new bottleneck arises. The idea of using the popular shared bus to implement the communication medium is no longer acceptable, mainly due to its high contention. Therefore, MPSoCs require an interconnection network [3] to connect the processors, as shown in Figure 5.1. Interconnection networks are classified according to the way resources are connected to the nodes.

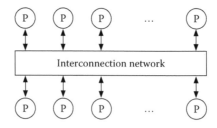

FIGURE 5.1　Interconnection network in a multiprocessor system.

Direct networks have resources connected directly to the nodes and use message passing for interprocess communication. Examples of topologies of this kind are linear, ring, mesh, cube, torus, and fully connected, as shown in Figure 5.2.

Indirect networks have resources connected at the end of the communication channels and use shared memory for interprocess communication. Examples of topologies for this kind of network are crossbar, as shown in Figure 5.3, and multistate, as shown in Figure 5.4, such as omega, baseline, butterfly, and cube [4,5].

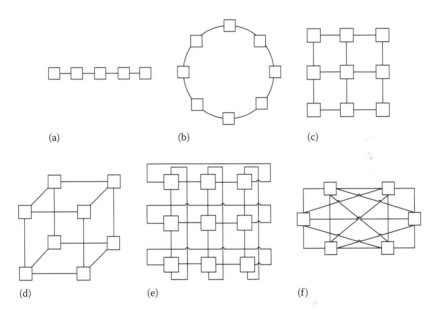

FIGURE 5.2   Direct network topologies: linear (a), ring (b), 2D mesh (c), cube (d), torus (e), and fully connected (f).

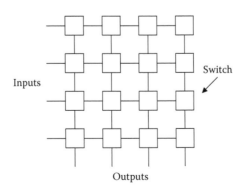

FIGURE 5.3   Crossbar topology.

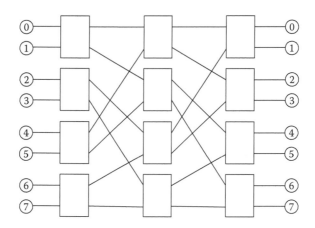

FIGURE 5.4   Multistage topology.

5.2 RELATED WORK

One of the first MPSoCs was Daytona [6], designed for wireless base stations. Daytona was a symmetric architecture with four CPUs attached to a high-speed split-transaction bus. The processors shared a common address space in memory.

Another example was the C-5 Network Processor [7], designed for packet processing in networks. Packets were handled by channel processors that were grouped into four clusters of four units each. Three buses handled different types of traffic in the processor.

An example of multimedia processor was the Philips Viper Nexperia [8], which included a microprocessor without interlocked pipeline stages (MIPS) and a Trimedia very long instruction word processor. The former acted as a master, running the operating systems, and the latter acted as a slave, carrying out commands from the MIPS. The system included one bus for each CPU and one bus for the external memory interface, using bridges to connect the buses.

Early cell phone processors performed baseband operations, including both communication and multimedia. The OMAP 5912, from Texas Instruments [9], had an ARM9 and a TMS320C55x digital signal processor. The first one acted as a master and the second one acted as a slave, performing signal processing operations.

Another example of cell phone application was the STMicroelectronics Nomadik [10], which used an ARM926EJ as its host processor, using a bus-based design.

The ARM MPCore [11] was a homogenous multiprocessor, also allowing some heterogeneous configurations. The architecture could accommodate up to four CPUs. The memory controller could be configured to offer different degrees of access to different parts of the memory for different CPUs. One CPU could be able to only read one part of the memory space, while other parts of the memory space could not be accessible to some CPUs.

The IXP2855 [12] was a network processor, in which 16 microengines were organized into two clusters to process packets. An XScale CPU served as host processor.

The Cell processor [13] had a PowerPC host and a set of eight processing elements known as *synergistic processing elements*. The processing elements, PowerPC, and I/O interfaces were connected by the element interconnect bus, which was built from four 16-B-wide rings. Each ring could handle up to three nonoverlapping data transfers at a time.

As MPSoCs grow in complexity, bus-based architecture will run out of performance and consume far more energy than desirable to achieve the required on-chip communications and bandwidth. This has led to the concept of network-on-chip (NoC) [14]. The idea is to use a hierarchical network with routers to allow packets to flow more efficiently between originators and targets, and to provide additional communications resources, rather than a single shared bus, so that multiple communications channels can be simultaneously operating. The NoC-based onchip interconnect is quite adequate for a heterogeneous multiprocessing system. In a homogenous multiprocessing system, the main idea is to have an application running in parallel, by distributing threads among the processors. In this kind of implementation, a shared memory is more adequate, allowing the processes to exchange information through shared variables. A good survey on MPSoC can be found in [15].

5.3 CROSSBAR TOPOLOGY

The crossbar network allows for any processor to access any memory module simultaneously, as far as the memory module is free. Arbitration is required when at least two processors attempt to access the same memory module. However, contention is not an usual case, happening only when processors share the same memory resource, for example, in order to exchange information. In this work, we consider a distributed arbitration control, shared among the switches connected to the same memory module. In Figure 5.5, the main components are introduced, labeled according to

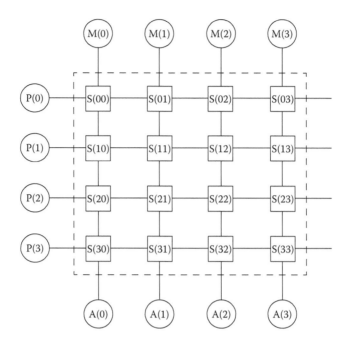

FIGURE 5.5 Crossbar components.

their relative position in the network, where i identifies the row and j identifies the column. For instance, component $A(j)$ corresponds to the arbiter [16] for column j. For the sake of legibility, we consider four processors ($0 \leq i \leq 3$) and four memory modules ($0 \leq j \leq 3$).

5.3.1 Processor

The processor is based on the PLASMA CPU core, designated MLite_CPU (MIPS Lite Central Processor Unit) [17], shown in Figure 5.6. In order to access a memory module $M(j)$, processor $P(i)$ must request the corresponding bus $B(j)$ and wait for the response from the arbiter $A(j)$. A bus access control is, then, implemented, as depicted in Figure 5.7, which decodifies the two most significant bits of the current address (ADDRESS(i)<29:28>), generating the corresponding memory module bus request REQ_IN(i, j). If $P(i)$ is requesting its primary bus $B(j)$, for which $i = j$, the arbiter sets signal PAUSE_CONT(i), pausing the processor until arbitration is complete, when, then, the arbiter resets this signal. On the other hand, if $P(i)$ is requesting a secondary bus $B(j)$, for which $i \neq j$, signal BUS_REQ(i) is set, pausing the processor until arbitration is complete, when, then, the arbiter sets signal DIS_EN(i), activating the processor.

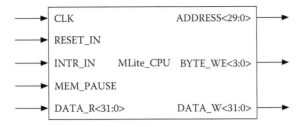

FIGURE 5.6   MLite_CPU interface signals.

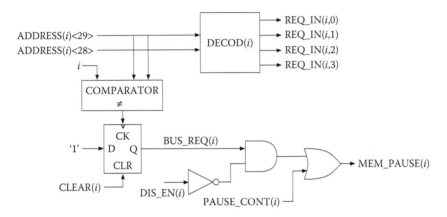

FIGURE 5.7   Bus access control.

5.3.2 Switch

The croosbar switch is basically a set of tri-state gates, controlled by the arbiter, as shown in Figure 5.8. Signal COM_DISC(i, j) is set by the arbiter, after arbitration is complete, establishing the communication between processor P(i) and memory module M(j).

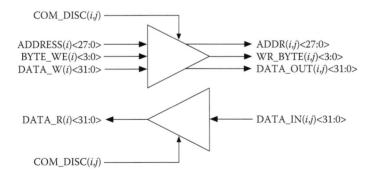

FIGURE 5.8   Crossbar switch.

5.3.3 Network Controller

The network controller is composed of the arbiter $A(j)$ and a set of controllers, one for each processor, implemented by state machines $SM(j)<0:N-1>$, as shown in Figure 5.9. Upon receiving a bus request, through signals $REQ\_IN(i, j)<0:N-1>$, the arbiter $A(j)$ schedules a processor to be the next bus master, based on the round-robin algorithm, by activating the corresponding signal $GRANT(i, j)$. State machines are used to control the necessary sequence of events to transition from the present bus master to the next one.

There are two types of state machines: primary and secondary. A primary state machine, designated $SM\_P(j)$, controls processor $P(i)$ bus accesses to its primary bus $B(j)$, for which $i = j$. A secondary state machine, designated $SM\_S(i, j)$, controls processor $P(i)$ bus accesses to a secondary bus $B(j)$, for which $i \neq j$. Therefore, for each arbiter, there will be one primary state machine and $N - 1$ secondary state machines.

5.3.3.1 Primary State Machine

The primary state machine is described by Algorithm 5.1. State **Reset** is entered whenever signal RESET goes to 1, setting signal PAUSE_CONT(i), which suspends $P(i)$, and resetting all the others. As RESET goes to 0, $SM\_P(j)$ enters state **Con_1**, establishing the connection of $P(i)$ to $B(j)$, for $i = j$, by setting signal CON_DISC(i, j), according to Figure 5.8. Once in state **Cont**, $P(i)$ starts the bus access, as signal PAUSE_CONT(i) goes to 0. While there are no requests from other processors, so $GRANT(i, j) = 1$ for

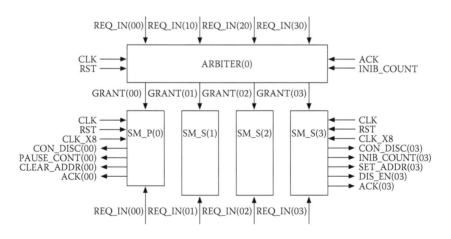

FIGURE 5.9   Network controller.

ALGORITHM 5.1   PRIMARY STATE MACHINE

```
  Reset: PAUSE_CONT ← 1;  CON _DISC ← 0; ACK_ME ← 0;
         CLEAR ← 0;
         if RESET = 1 then go to Reset;
  Con_1: PAUSE_CONT ← 1;  CON_DISC ← 1; ACK_ME ← 0;
         CLEAR ← 0;
         if CLK = 0 then go to Con_1;
   Cont: PAUSE_CONT ← 0;  CON_DISC ← 1; ACK_ME ← 0;
         CLEAR ← 0;
         if GRANT = 1 and REQ = 0 then goto Wait;
         else if GRANT = 0 and REQ = 1 then go to Pause;
   Wait: PAUSE_CONT ← 0;  CON_DISC ← 1; ACK_ME ← 0;
         CLEAR ← 0;
         if GRANT = 1 then go to Ack else go to Disc;
    Ack: PAUSE_CONT ← 0;  CON_DISC ← 0; ACK_ME ← 1;
         CLEAR ← 0;
   Disc: PAUSE_CONT ← 0;  CON_DISC ← 0; ACK_ME ← 0;
         CLEAR ← 0;
         if GRANT = 0 then go to Disc;
  Con_2: PAUSE_CONT ← 0;  CON_DISC ← 1; ACK_ME ← 0;
         CLEAR ← 0;
         if CLK = 0 then go to Con_2;
  Clear: PAUSE_CONT ← 0;  CON_DISC ← 1; ACK_ME ← 1;
         CLEAR ← 1;
         go to Cont;
  Pause: PAUSE_CONT ← 1;  CON_DISC ← 1; ACK_ME ← 0;
         CLEAR ← 0;
 P_Disc: PAUSE_CONT ← 1;  CON_DISC ← 0; ACK_ME ← 0;
         CLEAR ← 0;
         if GRANT = 0 then go to P_Disc else go to CON_1;
```

$i = j$, and P(i) is not requesting any other B(j), so REQ(i, j) = 1 for $i = j$, the primary state machine stays in state **Cont**. If another processor requests B(j), for $i = j$, then the arbiter resets signal GRANT(i, j), for $i = j$, and the primary state machine enters state **Pause**, in order to suspend P(i), by setting signal PAUSE_CONT(i). Next, SM_P(j) enters state **P_Disc**, in order to disconnect P(i) from B(j), by resetting signal CON_DISC(i, j). It stays in this state until the arbiter gives B(j) back to P(i), by setting signal GRANT(i, j), for $i = j$. SM_P(j), then, returns to state **Con_1**, where P(i) reestablishes its connection to B(j). On the other hand, from state **Cont**, the other possibility is that P(i) requests another B(j), for $i \neq j$, resetting signal REQ(i, j), for $i = j$. In this case, SM_P(j) enters state **Wait**, in order to check if the arbiter has already granted the secondary bus to P(i), in which case signal GRANT(i, j), for $i = j$, goes to 0, or not yet, in which case signal

GRANT(i, j), for $i = j$, remains in 1. In the first situation, SM_P(j) enters state **Disc**, in order to disconnect P(i) from B(j), for $i = j$, by resetting signal CON_DISC(i, j). In the second situation, SM_P(j) enters state **Ack**, setting signal Ack(j), in order to force the arbiter to reset signal GRANT(i, j), as shown in Figure 5.9. Once in state **Disc**, SM_P(j) stays in this state until the arbiter grants once again B(j) to P(i), for $i = j$, meaning that P(i) is now requesting access to its primary bus. As signal GRANT(i, j) goes to 1, SM_P(j) enters state **Con_2**, where P(i) is then connected to its primary bus, as signal CON_DISC(i, j) goes to 1. Then, SM_P(j) enters state **Clear**, in order to reset signal BUS_REQ, according to Figure 5.7, which was set when P(i) was addressing a secondary memory module, for which $i \neq j$. This signal, when set, pauses P(i), until the secondary state machine sends the control for P(i) to access the secondary bus.

5.3.3.2 Secondary State Machine

The secondary state machine is described by Algorithm 5.2. During initialization, when signal Reset is 1, SM_S(i, j) stays in state **Reset** until the arbiter grants a secondary bus B(j) to processor P(i). When signal GRANT(i, j) goes to 1, SM_S(i, j) enters state **Wait_1**, followed by state **Wait_2**, in order to give time to the corresponding primary state machine

---

ALGORITHM 5.2   SECONDARY STATE MACHINE

```
 Reset:  CON_DISC ← 0; DIS_EN ← 0; INIB_COUNT ← 0;
         ACK_ME ← 0;
         if GRANT = 0 then go to Reset;
Wait_1:  CON_DISC ← 0; DIS_EN ← 0; INIB_COUNT ← 1;
         ACK_ME ← 0;
Wait_2:  CON_DISC ← 0; DIS_EN ← 0; INIB_COUNT ←;
         ACK_ME ← 0;
   Con:  CON_DISC ← 1; DIS_EN ← 0; INIB_COUNT ← 1;
         ACK_ME ← 0;
         if CLK = 0 then go to Con;
   Dis:  CON_DISC ← 1; DIS_EN ← 1; INIB_COUNT ← 1;
         ACK_ME ← 0;
         if CLK = 0 then go to Dis;
    En:  CON_DISC ← 1; DIS_EN ← 0; INIB_COUNT ← 1;
         ACK_ME ← 0;
  Disc:  CON_DISC ← 0; DIS_EN ← 0; INIB_COUNT ← 1;
         ACK_ME ← 0;
   Ack:  CON_DISC ← 0; DIS_EN ← 0; INIB_COUNT ← 1;
         ACK_ME ← 1;
         go to Reset;
```

---

to pause $P(i)$ and disconnect it from $B(j)$, for $i = j$. In this case, either $P(i)$ is requesting a secondary bus or another processor is requesting $B(j)$ as secondary bus. In the first situation, SM_P(i, j) enters state **Wait** and in the second situation SM_P(i, j) enters state **Pause**, as described above. Observe that only the corresponding signal GRANT(i, j) is set, according to $P(i)$ and $B(j)$ in question. Next, SM_S(i, j) enters state **Con**, where signal CON_DISC(i, j) is set, connecting $P(i)$ to $B(j)$. Once in state **Dis**, signal DIS_EN(i, j) goes to 1, activating $P(i)$, as shown in Figure 5.7. Recall that $P(i)$, for $i = j$, was paused by the primary state machine, either because it requested a secondary bus or its primary bus is being requested by another processor. Once $P(i)$ finishes using $B(j)$, for which $i \neq j$, SM_S(i, j) enters state **En**, resetting signal DIS_EN(i, j) and pausing $P(i)$. Next, SM_S(i, j) enters state **Disc**, resetting signal CON_DISC(i, j) and disconnecting $P(i)$ from $B(j)$. Then, SM_S(i, j) enters state **Ack**, in order to tell the arbiter it finished using $B(j)$, by setting signal ACK_ME, which makes the arbiter select the next bus master. Observe that signal INIB_COUNT goes to 1 as soon as SM_S(i, j) leaves state **Reset**, stopping the counter that controls the time limit for $P(i)$ to use $B(j)$, for $i = j$, since this processor is not using its primary bus.

5.4 EXPERIMENTAL RESULTS

In order to analyze the performance of the proposed architecture, we used the particle swarm optimization (PSO) method [18,19] to optimize an objective function. This method was chosen due to its intensive computation, being a strong candidate for parallelization. In this method, particles of a swarm are distributed among the processors, and at the end of each iteration, a processor accesses the memory module of another one in order to obtain the best position found in the swarm. The communication between processors is based on three strategies: ring, neighborhood, and broadcast.

5.4.1 Particle Swarm Optimization

The PSO method keeps a swarm of particles, where each one represents a potential solution for a given problem. These particles transit in a search space, where solutions for the problem can be found. Each particle tends to be attracted to the search space, where the best solutions were found. The position of each particle is updated by the velocity factor $v_i(t)$, according to Equation 5.1:

$$x_i(t+1) = x_i(t) + v_i(t+1) \tag{5.1}$$

Each particle has its own velocity, which drives the optimization process, leading the particle through the search space. This velocity depends on its performance, called cognitive component, and on the exchange of information with its neighborhood, called social component. The cognitive component quantifies the performance of particle i, in relation to its performance in previous iterations. This component is proportional to the distance between the best position found by the particle, called $Pbest_i$, and its actual position. The social component quantifies the performance of particle i in relation to its neighborhood. This component is proportional to the distance between the best position found by the swarm, called $Gbest_i$, and its actual position. In Equation 5.2, we have the definition of the actual velocity in terms of the cognitive and social components of the particle:

$$v_i(t+1) = v_j(t) \times w(t) + c_1 \times r_1 \left[Pbest_i - x_i(t) \right]$$
$$+ c_2 \times r_2 \left[Gbest_i - x_i(t) \right] \tag{5.2}$$

Components r_1 and r_2 control the randomness of the algorithm. Components c_1 and c_2 are called the cognitive and social coefficients, controlling the trust of the cognitive and social components of the particle. Most of the applications use $c_1 = c_2$, making both components to coexist in harmony. If $c_1 \gg c_2$, then we have an excessive movement of the particle, making difficult the convergence. If $c_2 \gg c_1$, then we could have a premature convergence, making easy the convergence to a local minimum.

Component w is called the inertia coefficient and defines how the previous velocity of the particle will influence the actual one. The value of this factor is important for the convergence of the PSO. A low value of w promotes a local exploration of the particle. On the other side, a high value promotes a global exploration of the space. In general, we use values near to 1, but not too close to 0. Values of w greater than 1 provide a high acceleration to the particle, which can make convergence difficult. Values of w near 0 can make the search slower, yielding an unnecessary computational cost. An alternative is to update the value of w at each iteration, according to Equation 5.3, where n_{ite} is the total number of iterations. At the beginning of the iterations, we have $w \approx 1$, increasing the exploratory characteristic of the algorithm. During iterations, we linearly decrease w, making the algorithm to implement a more refined search.

ALGORITHM 5.3   PSO

```
Create and initialize a swarm with n particles;
repeat
  for i = 1 → n do
      Calculate the fitness of particle_i;
      if Fitness_i ≤ Pbest then
          Update Pbest with the new position;
      end
      if Pbest_i ≤ Gbest_i then
       Update Gbest_i with the new position;
      end
      Update the particle's velocity;
      Update the particle's position;
  end
until Stop criteria = true;
```

$$w(t+1) = w(t) - \frac{w(0)}{n_{ite}} \qquad (5.3)$$

The size of the swarm and the number of iterations are other parameters of the PSO. The first one is the number of existing particles. A high number of particles allows for more parts of the search space to be verified at each iteration, which allows for better solutions to be found, if compared with solutions found in smaller swarms. However, this increases the computational cost, with the increase in execution time. The number of iterations depends on the problem. With few iterations, the algorithm could finish too early, without providing an acceptable solution. On the other hand, with a high number of iterations, the computational cost could be unnecessarily high. Algorithm 5.3 describes the PSO method.

5.4.2 Communication between Processes

The parallel execution of the PSO method was done by allocating one instance of the algorithm to each processor of the network. The swarm was then equally divided among the processors. Each subswarm evolves independently and, periodically, *Gbest* is exchanged among the processors. This exchange of data was done based on three strategies: ring, neighborhood, and broadcast.

Figure 5.10 describes the ring strategy, while Algorithm 5.4 describes the PSO using this strategy for process communication. The neighborhood strategy can be depicted by Figure 5.11 and the PSO algorithm that implements this strategy is described by Algorithm 5.5. Figure 5.12 shows

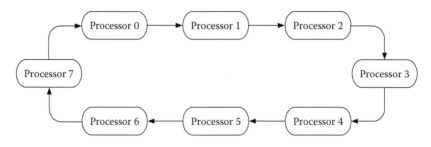

FIGURE 5.10   Ring strategy.

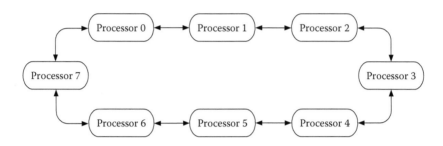

FIGURE 5.11   Neighborhood strategy.

ALGORITHM 5.4   PSO WITH RING STRATEGY

```
Create and initialize a swarm with n particles;
id:= processoridentification;
tmpid:= id - 1;
nproc:= numberofprocessorsinthenetwork;
if id ≠ 0 then
    endprocess(id):= 0;
end
tmpid:= id - 1;
repeat
    for j = 1→ n do
        Calculate the fitness of particleᵢ;
        Update Gbest(id) and Pbest(id);
        Update the particle's velocity;
        Update the particle's position;
    end
    Copy Gbest(id) to share the memory;
    Read Gbest from processor(tmpid);
    if Gbest(tmpid) ≤ Gbest(id) then
        Gbest(id):= Gbest(tmpid);
    end
until Stop criteria = true;
```

(Continued)

ALGORITHM 5.4 (Continued)   PSO WITH RING STRATEGY

```
if id = 0 then
    Best:= Gbest(id);
    tmpid:= id + 1;
    for k = 1 → nproc - 1 do
        Read endprocess(tmpid);
        while endprocess(tmpid) = 0 do
          Read endprocess(tmpid);
        end
        Read Gbest from processor(tmpid);
        if Gbest(tmpid) ≤ Best then
          Best:= Gbest(tmpid);
          end
          tmpid:= tmpid - 1;
    end
else
    endprocess(id):= 1;
end
```

ALGORITHM 5.5   PSO WITH NEIGHBORHOOD STRATEGY

```
Create and initialize a swarm with n particles;
id:= processoridentification;
tmpid:= id - 1;
nproc:= numberofprocessorsinthenetwork;
if id ≠ 0 then
    endprocess(id):= 0;
end
repeat
    for j = 1 → n do
        Calculate the fitness of particle_j;
        Update Gbest(id) and Pbest(id);
        Update the particle's velocity;
        Update the particle's position;
    end
    Copy Gbest(id) to share the memory;
    tmpid:= id + 1;
    Read Gbest from processor(tmpid);
    if Gbest(tmpid) ≤ Gbest(id) then
        Gbest(id):= Gbest(tmpid);
    end
    tmpid:= id - 1;
    Read Gbest from processor(tmpid);
```

(Continued)

ALGORITHM 5.5 (Continued)   PSO WITH NEIGHBORHOOD STRATEGY

```
      if Gbest(tmpid) ≤ Gbest(id) then
          Gbest(id) := Gbest(tmpid);
   end
  until Stop criteria = true;
  if id = 0 then
      Best:= Gbest(id);
      tmpid:= id + 1;
      for k = 1 → nproc - 1 do
       Read endprocess(tmpid);
       while endprocess(tmpid) = 0 do
          Read endprocess(tmpid);
       end
       Read Gbest from processor(tmpid);
       if Gbest(tmpid) ≤ Best then
          Best:= Gbest(tmpid);
       end
       tmpid:= tmpid + 1;
      end
  else
      endprocess(id) := 1;
  end
```

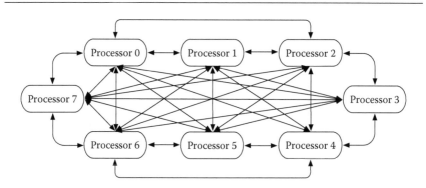

FIGURE 5.12   Broadcast strategy.

the broadcast strategy and Algorithm 5.6 describes its use for process communication by the PSO algorithm.

5.4.3 Performance Figures

For performance evaluation, we used the speedup defined by Amdahl law and expressed by Equation 5.4:

$$S = \frac{T_1}{T_2} \tag{5.4}$$

ALGORITHM 5.6   PSO WITH BROADCAST STRATEGY

```
Create and initialize a swarm with n particles;
id:= processoridentification;
nproc:= numberofprocessorsinthenetwork;
if id ≠ 0 then
    endprocess(id):= 0;
end
repeat
  for j = 1 → n do
      Calculate the fitness of particle_j;
      Update Gbest(id) and Pbest(id);
      Update the particle's velocity;
      Update the particle's position;
  end
  Copy Gbest(id) to share the memory;
  tmpid:= id + 1;
  for k = 1 → nproc - 1 do
     Read Gbest from processor(tmpid);
     if tmpid = nproc - 1 then
         tmpid = 0;
     else
         tmpid:= tmpid + 1;
     end
  end
until Stop criteria = true;
if id = 0 then
    Best:= Gbest(id);
    tmpid:= id + 1;
    for k = 1 → nproc - 1 do
        Read endprocess(tmpid);
        while endprocess(tmpid) = 0 do
           Read endprocess(tmpid);
        end
        Read Gbest from processor(tmpid);
        if Gbest(tmpid) ≤ Best then
           Best:= Gbest(tmpid);
        end
        tmpid:= tmpid + 1;
    end
else
    endprocess(id):= 1;
end
```

where T_1 is the execution time of an application using one processor and T_2 is the execution time using multiple processors. Efficiency reflects the use of the N processors in a multiprocessor system. Therefore, the relationship between speedup and efficiency can be expressed by Equation 5.5:

$$E = \frac{S}{N} \tag{5.5}$$

For the first experiment, the PSO algorithm was used to minimize the Rosenbrock function, defined by Equation 5.6 and whose curve is shown in Figure 5.13. We used 1, 2, 4, 8, 16, and 32 processors for each simulation, considering each of the communication strategies, with a total of 64 particles, distributed among the processors, and running 32 iterations.

$$f(x, y) = 100 \left[y - \left(x^2 \right) \right]^2 + \left(1 - x \right)^2 \tag{5.6}$$

The simulation results obtained, using the ring strategy, are presented in Table 5.1, where NP means the number of processors, PP means particle

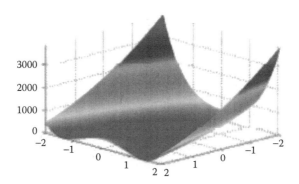

FIGURE 5.13   Graphic of the Rosenbrock function.

TABLE 5.1   Simulation Results for the Minimization of the Rosenbrock Function Using the Ring Strategy

NP	PP	CC	S	E
1	64	10232089	1.000000000	1.000000000
2	32	5151219	1.986343233	0.993171616
4	16	2582747	3.961707825	0.990426956
8	8	1304211	7.845424552	0.980678069
16	4	663325	15.42545359	0.964090849
32	2	343537	29.78453267	0.930766646

per processor, CC means clock cycles, S means speedup, and E means efficiency. When using the neighborhood strategy, the simulation results obtained are presented in Table 5.2. For the broadcast strategy, the simulation results obtained are shown in Table 5.3. The speedup obtained is described by Figure 5.14 and the efficiency is described by Figure 5.15.

As a second experiment, the PSO was used to minimize the Rastrigin function, described by Equation 5.7 and shown in Figure 5.16. As for the first experiment, we used 1, 2, 4, 8, 16, and 32 processors for each simulation, considering each of the communication strategies, with a total of 64 particles, distributed among the processors, and running 32 iterations. The simulation results obtained, when using the ring strategy, can be seen in Table 5.4. When using the neighborhood strategy, we obtained the simulation results presented in Table 5.5. For the broadcast strategy, the simulation results can be shown in Table 5.6. The speedup obtained is described by Figure 5.17 and the efficiency is described by Figure 5.18.

$$f(x, y) = 20 + x^2 + y^2 - 10\cos(2\pi x) - 10\cos(2\pi y) \qquad (5.7)$$

TABLE 5.2   Simulation Results for the Minimization of the Rosenbrock Function Using the Neighborhood Strategy

NP	PP	CC	S	E
1	64	10232089	1.000000000	1.000000000
2	32	5151219	1.986343233	0.993171616
4	16	2595968	3.941531252	0.985382813
8	8	1314996	7.781079942	0.972634993
16	4	671576	15.23593607	0.952246004
32	2	349769	29.25384754	0.914182736

TABLE 5.3   Simulation Results for the Minimization of the Rosenbrock Function Using the Broadcast Strategy

NP	PP	CC	S	E
1	64	10232089	1.000000000	1.000000000
2	32	5151219	1.986343233	0.993171616
4	16	2602982	3.930910394	0.982727599
8	8	1367434	7.482693132	0.935336641
16	4	801790	12.76155727	0.797597329
32	2	617331	16.57472085	0.517960027

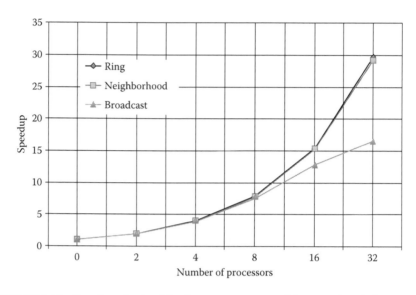

FIGURE 5.14   Speedup obtained for the minimization of the Rosenbrock function.

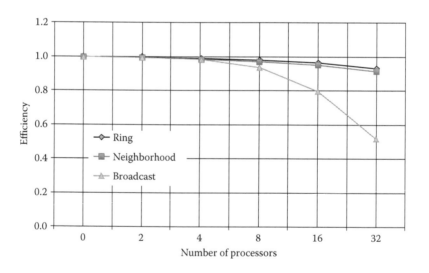

FIGURE 5.15   Efficiency obtained for the minimization of the Rosenbrock function.

5.5 CONCLUSIONS

In order to evaluate the performance offered by the proposed architecture, we executed the PSO method for the minimization of the Rosenbrock function and of the Rastrigin function, both sequentially and in parallel. The simulation was done for 1, 2, 4, 8, 16, and 32 processors, using a swarm

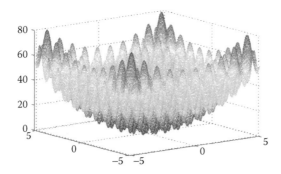

FIGURE 5.16   Graphic of the Rastrigin function.

TABLE 5.4   Simulation Results for the Minimization of the
Rastrigin Function Using Ring Strategy

NP	PP	CC	S	E
1	64	46166574	1,000000000	1,000000000
2	32	23088801	1,999522366	0,999761183
4	16	11575210	3,988400556	0,997100139
8	8	5799370	7,960618826	0,995077353
16	4	2942026	15,69210265	0,980756416
32	2	1485505	31,0780334	0,971188544

TABLE 5.5   Simulation Results for the Minimization of the
Rastrigin Function Using Neighborhood Strategy

NP	PP	CC	S	E
1	64	46166574	1,000000000	1,000000000
2	32	23088801	1,999522366	0,999761183
4	16	11575246	3,988388152	0,997097038
8	8	5810538	7,945318316	0,99316479
16	4	2952458	15,63665732	0,977291083
32	2	1498289	30,81286321	0,962901975

TABLE 5.6   Simulation Results for the Minimization of the
Rastrigin Function Using Broadcast Strategy

NP	PP	CC	S	E
1	64	46166574	1,000000000	1,000000000
2	32	23088801	1,999522366	0,999761183
4	16	11600572	3,97968083	0,994920207
8	8	5853974	7,886364716	0,985795589
16	4	3107786	14,85513288	0,928445805
32	2	1747628	26,41670539	0,825522043

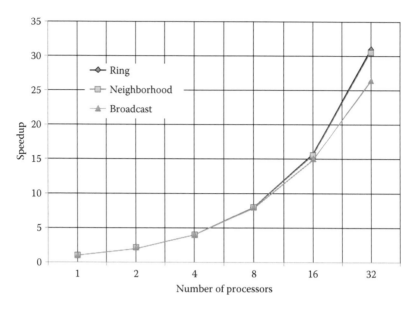

FIGURE 5.17   Speedup obtained for the minimization of the Rastrigin function.

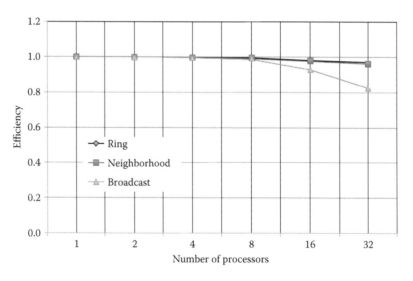

FIGURE 5.18   Efficiency obtained for the minimization of the Rastrigin function.

of 64 particles and running 32 iterations. We exploited three communication strategies: ring, neighborhood, and broadcast. The speedup obtained
demonstrated that the performance offered by the network increases with
the number of processors, demonstrating its scalability. Another fact is
that both ring and neighborhood strategies have similar impact on the

performance of the PSO execution, while the broadcast strategy decreases the performance. This decrease is due to the fact that the latter imposes much more interprocess communication than the former ones.

As for future work, we intend to explore other applications for parallelization, in order to analyze the impact of their behavior, specially concerning the interprocess communication. Also, we intend to introduce cache memory, to improve performance, having foreseen the problem of cache coherency. Another aim is to develop a microkernel, to implement task scheduling and explore multithread execution. The network controller, necessary for each of the memory modules, consisting of the arbiter and a number of state machines proportional to the number of processors impacts both the execution time and the area required. Therefore, it is our aim to investigate other options in search for better solutions. Since all the results were based on simulation, our next step is to synthesize the architecture, in order to analyze the cost × performance relation.

ACKNOWLEDGMENTS

We would like to thank the Brazilian federal agency, CNPq, and the Rio de Janeiro state agency, FAPERJ, for their financial support.

REFERENCES

1. A. S. Tanenbaum. *Structured Computer Organization*. Pearson/Prentice Hall, Upper Saddle River, NJ, 2006.
2. D. A. Patterson and J. L. Hennessy. *Computer Organization: The Hardware/Software Interface*. Morgan Kaufmann, San Francisco, CA, 2005.
3. L. M. Ni. Issues in designing truly scalable interconnection networks. In *Proceedings of the International Conference on Parallel Processing*, pp. 74–83. IEEE Press, 1996.
4. J. Duato, S. Yalamanchili, and L. Ni. *Interconnection Networks: An Engineering Approach*. Morgan Kaufmann, San Francisco, CA, 2003.
5. P. T. Pande et al. Design, synthesis, and test of networks on chips. *IEEE Des. Test Comput.*, 22(5):404–413, 2005.
6. B. Ackland et al. A single-chip, 1.6-billion, 16-b mac/s multiprocessor dsp. *IEEE J. Solid-State Circ.*, 35(3):412–424, 2000.
7. *C-5 Network Processor Architecture Guide*, C-Port Corporation, Andover, MA, May 2001.
8. S. Dutta, R. Jensen, and A. Rieckmann. Viper: A multiprocessor soc for advanced set-top box and digital tv systems. *IEEE Des. Test Comput. Test Comput.*, 18(5):21–31, 2001.
9. *Omap5912 Multimedia Processor Device Overview and Architecture, Reference Guide*, http://www.ti.com, Dallas, TX, 2004.

10. A. Artieri et al. *Nomadik-Open Multimedia Platform for Next Generation Mobile Devices*, STMicroelectronics, Chemin du Champ des FillesPlan-Les_Ouates, Geneva, CH, 2003.

11. J. Goodacre and A. N. Sloss. Parallelism and the arm instrucion set architecture. *Computer*, 38(7):42–50, 2005.

12. Intel Corportation. Intel ixp2855 network processor, Santa Clara, CA, 2005.

13. M. Kistler, M. Perrone, and F. Petrini. Cell multiprocessor communication network: Built for speed. *IEEE Micro.*, 26(3):10–23, May 2006.

14. G. De Micheli and L. Benini. *Networks on Chips: Technology and Tools*. Morgan Kaufmann, San Francisco, CA, 2006.

15. W. Wolf, A. A. Jerraya, and G. Martin. Multiprocessor system-on-chip (mpsoc) technology. *IEEE Trans. Comput. Aided Des. Int. Circ. Syst.*, 27(10):1701–1713, 2008.

16. M. Weber. *Arbiters: Design Ideas and Coding Styles*. Technical report, Silicon Logic Engineering, Eau Claire, WI, 2001.

17. Plasma – most MIPS I(TM) opcodes: Overview, http://www.opencores.org, 2013.

18. J. Kennedy and R. Eberhart. Particle swarm optimization. In *Proceedings of the IEEE International Conference on Neural Networks*, volume 4, pp. 1942–1948. IEEE Press, Los Alamitos, CA, 1995.

19. A. P. Engelbrecht. *Fundamentals of Computational Swarm Intelligence*. John Wiley & Sons, Hoboken, NJ, 2006.

CHAPTER 6

Extended Quality of Service Modeling Based on Multiapplication Environment in Network-on-Chip

Abdelkader Saadaoui and Salem Nasri

CONTENTS

6.1 INTRODUCTION

Network-on-chip (NoC), as a new system paradigm for communications implemented on a chip, is useful by handling parallelism, manufacturing complexity, wiring problems, and reliability. It has tackled the system-on-chip (SoC) disadvantages (Al Faruque et al., 2010; Benini and De Micheli, 2002; De Micheli et al., 2010; Yiping et al., 2010).

NoCs are emerging as an attractive solution for the problem of the existing interconnection constraints implementing future networks (Bolotin et al., 2004; Goossens et al., 2005; Rijpkema et al., 2001, 2003).

NoC researchers have integrated techniques like routing and packet-switching concepts of computer networks into a chip (Bjerregaard and Mahadevan, 2006).

NoC is composed of intellectual property (IP) cores such as routers and network interfaces, connected among themselves by communicating channels (Rijpkema et al., 2003).

In wormhole switching, every packet is composed of the header, payload, and trailer. It is divided into one or more flits (flow digit) (Kim et al., 2005; Murali et al., 2006). A phit (physical unit) corresponds to the quantity of bits that can be transported in one time on the link.

Currently, applications need more performances in direct link with the NoC architecture, so differentiated services are provided through either class of services (CoS) or quality of services (QoS).

Researchers present CoS as a way of managing a group of similar types of traffic and treating each type as a class with its own level of service priority, otherwise as the classification of a specific traffic. While QoS involves guaranteeing service levels (SLs) to traffic flows, it specifies a guaranteed parameter level. In other words, QoS measures the performance degree in a data transfer system.

QoS is estimated through its parameters as well as in grid environment (Chunlin and Layuan, 2008; Plestys et al., 2007; Zhiang and Junzhou, 2006). However, several attempts were made to find a quantifiable scale for QoS measurement.

In this work, we address the QoS metrics problem for NoC-based system. We propose an extended approach of QoS metrics modeling and analysis based on dynamic routing for multiapplication environment with multiparameters.

Therefore, researchers are looking for a projection of QoS on quantifiable space, since it is qualitative, subjective, and not measurable.

Because of the weakness of a standard method for QoS estimation, this chapter proposes a tentative QoS evaluation using parameter values, in which we succeeded in assigning a quantifiable representation by making it through an estimation of other quantities that influence the QoS.

The remaining part of this chapter is structured as follows: Section 6.2 discusses related works; then, Section 6.3 explains the NoC target architecture. Section 6.4 presents dynamic routing techniques. Section 6.5 focuses on QoS metrics modeling requirements. In Section 6.6, we present the experimentation results and analysis. Finally, we conclude and look forward to future work in Section 6.7.

6.2 RELATED WORKS

NoC studies have used techniques of computer networks into a chip. In 2000, SPIN (LIP6) constituted the first study of NoC packets commutation (Bjerregaard and Mahadevan, 2006). After 2002, several research groups have focused on parameters as bandwidth and latency to ensure the QoS performance in the NoC.

QoS has received broad attention, so many techniques have been used to provide its definition; however, QoS should not be confused with CoS. In other words, CoS does not offer guarantees with bandwidth or delivery time since it is based on a best effort basis.

Lima et al. (2004, 2005) proposed a distributed admission control model that provided a uniform solution for managing the quality of multiple services in CoS-IP networks. Also, they extended the proposed model for a multiclass and multidomain environment (Lima et al., 2004, 2005). Bolotin et al. (2004), in the quality of service NoC (QNoC) project, identified four classes of service into SoC intermodule communication and defined an appropriately split SL. The four classes are presented in highest to lowest priority on the NoC architecture as: signaling, real time, RD/WR, and block transfer, successively for intermodule control signals, delay-constrained bit stream, short data access, and data blocks. QNoC design reinforces traffic flow by increasing network performance through low latency and high throughput (Thp) (Bolotin et al., 2004).

In parallel, some researches that studied QoS metrics for NoC inspired from macronetwork referred QoS as the capability of a network to provide better service.

Beigne et al. (2005) proposed a QoS integrated in complete asynchronous NoC architecture. This architecture targets globally asynchronous locally synchronous SoC. This provides low latency service using virtual channels (VCs) (Beigne et al., 2005). However, in Goossens et al. (2005) and Marescaux and Corporaal (2007), the authors combined best effort and guaranteed Thp services to ensure QoS NoC. Nevertheless, more software development is needed to further expose QoS features, such as proposed by Joven et al. (2010); they considered an integrated hardware–software approach for delivering QoS at the application level for NoC-based platforms (Joven et al., 2010). On the other hand, due to the necessity to define QoS in this promising interconnect paradigm, Bjerregaard and Mahadevan (2006) presented QoS as service quantification to the demanding core offered by NoC (Bjerregaard and Mahadevan, 2006). Yang and Huacan (2007) defined the QoS as a manager of distance between a service provider and a service requester and have normalized QoS parameters. They proposed an algorithm to normalize parameter values (Yang and Huacan, 2007).

In addition, Helali et al. (2005, 2006) and Helali and Nasri (2009) addressed the problem of metrics for end-to-end QoS management on real-time applications by presenting a virtual communication support (Helali et al., 2005). Their research was focused on the study of QoS through the switch buffering requirements (Helali et al., 2006). Helali and Nasri (2009) were interested in NoC switch scheduling and its impact on QoS metrics. Recently, Liu et al. (2012) proposed a novel agent-assisted QoS-based routing algorithm for wireless sensor network applications. The proposed algorithm shows that it can ensure better QoS by improving network performance such as delay, bandwidth, and packet loss parameters to increase the QoS level of network.

To our knowledge, few people are working on the same subject; in fact, Nasri (2011a) proposed a new approach of QoS metrics modeling based on the QoS parameter estimation and applications priority. We argue that implementing a quantifiable representation of QoS can be used to provide a NoC services arbiter. Thus, the main contribution leads to quantifiable representation of QoS; we are proposing a new approach for QoS evaluation and measurements. We think that the idea described in this chapter will motivate research in the promising future.

6.3 NoC STUDY

6.3.1 NoC Topology

The topology designates the structure of the connections of several cores of the NoC, in order to ensure a data exchange (Benini and De Micheli, 2006; Dally and Towles, 2001; Tenhunen, 2003; Winter et al., 2010). We propose a 4 × 4 mesh topology as shown in Figure 6.1.

The studied NoC architecture assumes that each router has a set of bidirectional ports linked to its neighbor routers and an IP core. Each router has five bidirectional ports: east, west, north, south, and local. The local port is used to connect its IP core via network interface. The other ports are connected to the neighbor routers. Each router has two (L2), three (L3), or four (L4) bidirectional links with neighbors, depending on the position of each one in the graph (Figure 6.3). In this study, we considered three different sinks connected to router 33 (L2), router 32 (L3), and router 22 (L-). High-quality multimedia communications over an NoC designed for data communications are a complex challenge. In fact, different application traffic characteristics are expected to be identical with real network scenarios like constant bit rate (CBR), variable bit rate (VBR), and transmission control protocol (TCP) traffics. It can be sent from

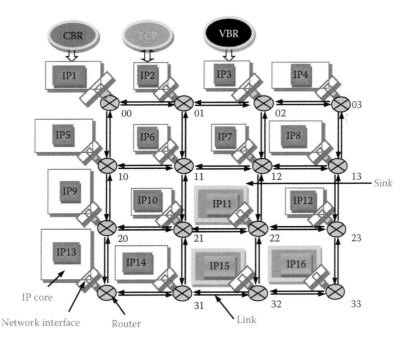

FIGURE 6.1   4 × 4 mesh NoC structure.

sources to destinations, generally used to simulate multidata traffic such as multimedia streaming systems.

CBR is an encoding method that keeps the bit rate the same as opposed to VBR that varies the bit rate. However, TCP ensures that all data arrive accurately and intact at the destination by sending acknowledgments on proper transfer or loss of packets. It requires handshaking to set up end-to-end communications, which means that links are bidirectional in order for acknowledgments to return to the source.

CBR and VBR traffics use user datagram protocol (UDP) as the transport layer. UDP offers better speed. It is commonly used for streaming applications that require fast transmission of data. It is faster than TCP because there is no flow control or error correction. However, TCP is used in case of non-time-critical applications, providing reliable end-to-end communication. There is a guaranteed delivery. It guarantees that all sent packets will reach the destination in the correct order.

6.3.2 QoS–IP Definition

The current work presents an intrinsic QoS module for on-chip communication that offers different QoS parameters.

Sources ap_i send packets to a destination under a same traffic configuration, and as shown in Figure 6.2, a QoS–IP is attached to a network interface of a router. QoS module collects data from the entire NOC, including the information necessary to calculate performance metrics, such as packet loss, latencies, and Thp, and analyses and injects decision into NoC router ports.

It is said that these decisions are part of a communication flow. A QoS–IP can inject more than one communication flow into the network, for a same or for different router ports.

QoS–NoC inputs/outputs are used to improve the resource utilization for NoC applications.

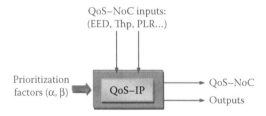

FIGURE 6.2 QoS–IP module.

QoS guarantees are important if the network capacity is insufficient; however, the QoS–NoC inputted in Figure 6.2, used as a quality measure, delivers NoC state to QoS–IP module. The acquired QoS information includes data traffic parameters such as end-to-end delay (EED), Thp, and packet loss rate (PLR), for describing a flow of multiapplication (CBR, VBR, TCP, etc.) environment in NoC, whereas prioritization factors of parameter (α) and application (β) values impact several QoS parameters, which are used by QoS–IP module.

Prioritization factors approach enables NoC environment to provide better service to certain flows, which is done by either raising the priority of a flow or limiting the priority of another flow (used later in QoS modeling). Also important is making sure that providing priority for one or more flows does not make other flows fail.

By implementing the QoS model, QoS–IP module controls router ports load within NoC. It has the ability to provide different priority to different application flows or to guarantee a certain level of performance to a data flow. QoS–IP module tends to make a system self-control. We intend to implement a feedback control QoS algorithm to achieve both high availability and load balancing on the NoC.

The QoS–IP in NoC architecture (Figures 6.2 and 6.3) assists to improve network performance by allowing the capability of reserving routes

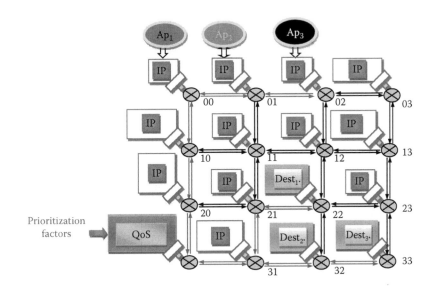

FIGURE 6.3   QoS data flow.

between sources–destinations and arbitrating flits to ensure end-to-end QoS parameters.

Each application ap_i generates packets that have different priority classes in terms of delivery probability. A flit transfer, which carries an SL priority index, can be classified by higher SL priority, and if channel is busy, QoS module can be intervening.

This study is motivated by avoiding many data transfer problems (deadlock, starvation, and drop) and improving flit deliveries of high priority class; the proposed scheduling policy focuses on improving flit deliveries of priority class. It presents a novel traffic management strategy supporting QoS for NoC by utilizing flit properties and integrated hardware–software QoS metrics modeling approach to exploring model support for NoC services arbiter.

6.3.3 Router Flow

Network congestion occurs when a link or router is carrying so much data that its QoS deteriorates. In Figure 6.3, many routers, like 22 and 32, have a routing problem (worst case) to forward a set of flits; an appropriate strategy is needed to ensure flit transfer and to prevent dropped data.

It is clear that if too many packets try to access a router, the load exceeds the router's capacity. As a result, the router will become overloaded with too much data, the input queue will fill up, and a set of packets will be dropped.

Figure 6.4 shows many packets moving from source to destination and links such as CBR, VBR, TCP with its ACK, and QoS flows. Traffic may be required to traverse the same link(s). We can also have a routing problem (worst case) to forward a set of flits. The NoC resources such as channels and buffers are limited. The QoS–IP module is used to deal with this kind of problem; its goal is to provide better service to certain flows. In other words, prioritizing traffic ensures that important traffic gets the fastest handling by giving all of the packets a weighting value based on priority. Once the packets are all given weights, they are transmitted according to the weight order (Das et al., 2009; Li et al., 2001; Nachiappan et al., 2012).

Because every packet is composed of many flits, we propose a solution used in router architecture. The solution proposes VCs within SLs for each router port to increase transfer quality of flits. Flit consists of QoS bits, flit-type bits, SL bits, VC bits, and data (Figure 6.5). Flit type indicates if it is a QoS flit, a header, a body, or a tail with:

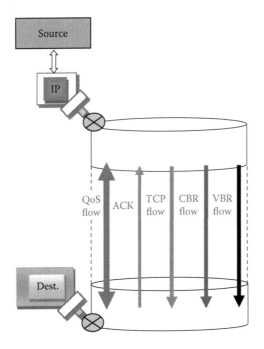

FIGURE 6.4   Flows in channel.

Flit type	QoS	SL	VC	Data

FIGURE 6.5   Flit format.

00: QoS flits

01: Packet header

10: Packet body

11: Packet tail

QoS bits uses in feedback control with QoS request, Busy_channel. The impact of multi-SLi with multi-VCs may be justified by increasing Thp and decreasing delay transfer flits. Thus, we assume that flit entering through router port carrying strategy idea does not drop or loop back. In Figure 6.6, classifier sets flits to the appropriate VC inside SL, while arbiter, in last router stage, selects a flit according to SL priority. Idea of

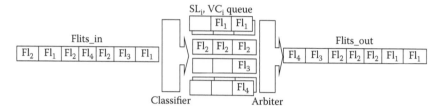

FIGURE 6.6   Router flit flow.

employing VC for each SLi will allow better utilization of router links and
improve a best communication delay, decreasing congestion.

6.4 DYNAMIC ROUTING TECHNIQUES

Routing algorithm's purpose is to define a scheme for transferring a packet
from source to destination (Yunus et al., 2011). Furthermore, the primary
goal of dynamic routing technique is to ensure that data transferred to the
network reach their destination.

The dynamic routing, used in this study, uses adaptive protocols to find
the shortest path between source and destination. It defines the faulty or
the broken path. This fault is declared in the entire network, and another
possible optimal route is chosen (Dally and Aoki, 1993).

The advantage of dynamic routing is its potential to improve performance and reduce congestion by choosing more optimal links.

For evaluation of our strategy performance, destinations are linked
with three types of routers. We select routers having two (L2), three (L3),
or four (L4) links for three different sources.

To meet ideal network behavior, we select randomly horizontal and/or
vertical failed path scenarios on the entire NoC, using random broking
links for a duration time (NoC state changes). This forces the system to
search a new path between sources and destination.

Three applications such as TCP, VBR, and CBR are concurrently active
in the same condition and in the same time. When horizontal and/or vertical links are broken, packets go dynamically through other router using
the shortest path to the destination. In Figure 6.7, *t* presents the iteration
time of simulation; however, duration time is the total simulation time.
These steps are realized in the flowchart in Figure 6.7.

Program flowchart begins by reading inputs such us topology dimension, transmission time, application characteristics, network, queuing, and
set-up routing and shows the sequence of instructions. Our approach is

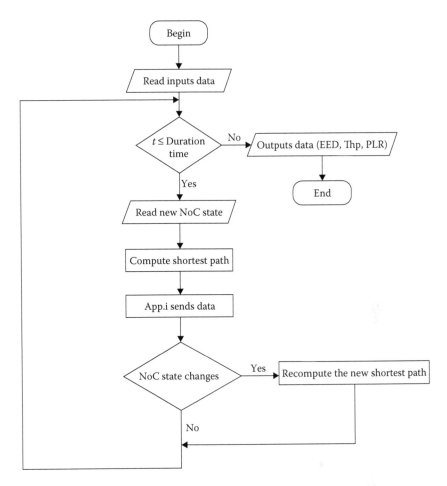

FIGURE 6.7   Flowchart of the dynamic routing scenarios.

presented by a subroutine, which is trying to meet a real environment with dynamic problem by creating a randomly failed link for a random time.

We used a dynamic routing algorithm based on distance vector (DV) for determining the shortest path among the available links by using the code ($ns rtproto DV). A dynamic routing table is updated periodically.

NoC has many horizontal and vertical links, and so to check our approach with a real NoC environment, we used scenarios of many routing problems such as broken or restored links for a random time by using the specific code ($ns rtmodel-at [random time] [down/up] [random {node_id} random {node_id + 1}]). In other words, breaks may be at arbitrary links for an arbitrary time; NS2 starts a new route discovery process and recomputes the best route with minimal expected travel time. It allows application nodes to obtain routes quickly for new destinations when links break/restore and to achieve a real network behavior. Our approach offers quick adaptation to dynamic link conditions and determines the shortest routes to destinations within the NoC.

An AWK script is used to analyze the output trace files. The script should calculate the QoS parameters such as EED, Thp, and PLR.

6.5 QoS METRICS MODELING REQUIREMENTS

6.5.1 QoS Definition

QoS, as an abstract qualitative notion, refers to defined measure of performance and levels of guarantees given in a data communications system.

For example, the user does not know details about video transmission but is satisfied if the video is received in the correct way without information loss.

To ensure such delivered application, a traffic contract is negotiated between the network application consumer and the provider. This contract ensures a minimum bandwidth along with the maximum delay that can be supported. Since there is no common or formal QoS metrics definition, we propose a new QoS metrics approach based on the prioritization factors and parameters. Each application needs different levels of performance.

Typically, QoS parameters include Thp, EED, jitter, and PLR. QoS parameters concern also the priority, reliability, speed, and amount of traffic sending over a network (Nasri, 2011a, b).

6.5.2 End-to-End Delay, Throughput, and Packet Loss Rate

EED refers to the time taken for a packet to be transmitted from source to destination. It includes the time elapsed in each node (source-routers) and on links through the communication path until the packet reaches its destination. The delay is unpredictable depending on the state of the network. It can be calculated by:

$$EED = \sum_{i=1}^{n} \left[\begin{array}{c} \text{Time of received packet}(i) - \\ \text{Time of transmitted packet}(i) \end{array} \right] \tag{6.1}$$

where:

n is the number of received packets

Thp defines how much useful data can be transmitted from source to destination per unit time. Thp is measured in bits per second (bps) and it is calculated by:

$$Thp = \frac{\sum_{i=1}^{n} \left\{ \left[\text{Time of received packet}(i) \right] \times 8 \right\}}{\text{Total duration of simulation}} \qquad (6.2)$$

where:

n is the received packet in byte

Packet loss happens when data packets are discarded in a network when a router is overloaded or tired down and cannot accept any more incoming data at a given moment.

Packet loss occurs when one or more packets of data traveling across a computer network fail to reach their destination. The packet loss or drop rate is the failure of one or more transmitted packets to arrive at their destination per the number of transmitted packets. It is calculated by:

$$PLR = \frac{\text{Number of transmitted packets} - \text{Number of received packets}}{\text{Number of transmitted packets}} \qquad (6.3)$$

6.5.3 QoS Modeling

Nasri (2011a, b) proposed an approach of QoS metrics model based on QoS parameter prioritization factors α_i for one application service using the relation:

$$QoS = \sum_{i=1}^{n} \alpha_i \times p_i \qquad (6.4)$$

where α_i is the prioritization factor of application, arbitrarily fixed referring to the following equations for one application p_i:

$$\sum_{i=1}^{n} \alpha_i = 1 \qquad (6.5)$$

However, in this chapter, we extended the approach of QoS modeling based on multiparameter and multiapplication environment in NoC.

In a multi-application environment $(ap_1, ap_2, ..., ap_m)$, we define for each application ap_i a set of parameters $(pi_1, pi_2, pi_3, ..., pi_n)$.

QoS performance parameters should be normalized as $p^{\wedge}_{ij}$, with $p_{ij\max}$ = Max$\{p_{ij}\}$ and $p_{ij\min}$ = Min$\{p_{ij}\}$ (Yang and Huacan, 2007). Then,

1. For increasing parameters when application value increases:

$$\hat{p}_{ij} = \left| \frac{p_{ij} - p_{ij\min}}{k \times p_{ij\max} - p_{ij\min}} \right| \tag{6.6}$$

2. For decreasing parameters when application value increases:

$$\hat{p}_{ij} = \left| \frac{p_{ij\max} - p_{ij}}{k \times p_{ij\max} - p_{ij\min}} \right| \tag{6.7}$$

3. $k \geq 1$ represents the network efficiency coefficient (e.g., in our case we chose $k = 1.03$).

If we suppose that we have m applications, QoS can be expressed by the following model:

$$ap_1 = \alpha_{11} \times \hat{p}_{11} + \alpha_{12} \times \hat{p}_{21} + \ldots + \alpha_1 n \times \hat{p}n_1 \tag{6.8}$$

$$ap_2 = \alpha_{21} \times \hat{p}_{12} + \alpha_{22} \times \hat{p}_{22} + \ldots + \alpha_2 n \times \hat{p}n_2 \tag{6.9}$$

$$ap_m = \alpha_{m1} \times \hat{p}_{1m} + \alpha_{m2} \times \hat{p}_{2m} + \ldots + a_{mn} \times \hat{p}_{nm} \tag{6.10}$$

Then,

$$\text{QoS} = \text{QoS}_0 + \beta_1 \times ap_1 + \beta_2 \times ap_2 + \ldots + \beta_m \times ap_m \tag{6.11}$$

6.6 EXPERIMENTATION, RESULTS, AND ANALYSIS

In our study, we consider different types of router interconnection depending on the position of the router on the NoC. The destinations 1, 2, and 3 are connected, respectively, to routers 22, 32, or 33, which have four (L4), three (L3), or two (L2) ports (Figure 6.2).

All links have a capacity of 100 Mbps. The traffic is identical for each source–sink pair. The on–off UDP sources send bursts of 500 packets during an on-period and have a 150 ms off-period. All TCP sources are greedy, that is, they always have data to transmit.

Each application sends data according to the used protocol (TCP and UDP). The use of drop tail algorithm, which is a first-in, first-out (FIFO) queue, is not fair between packet losses for applications. To improve the fairness between flows, we replaced the queue management of links using stochastic fairness queuing (SFQ) algorithm, which is supposed to be an equitable distribution algorithm.

Three applications CBR, VBR, and TCP are linked, respectively, to routers 00, 01, and 02. The communication of these applications starts simultaneously at the same time using a dynamic routing. We analyze some QoS metrics such as EED, Thp, and PLR in the NoC nodes.

These parameters are used to measure the QoS of an IP connection and quantify end-to-end NoC performance. NoC should ensure the negotiated QoS by satisfying certain values of these parameters. More details are given in Saadaoui and Nasri (2012).

6.6.1 End-to-End Delay

First, the average EED in the NoC is compared. Figure 6.8 shows the relationship between EED average and available packet size with router that has four (L4) bidirectional links with neighbors. Second, from Figure 6.8, we can observe that, contrary to CBR, VBR is the application that gives better results to network performances.

EED is affected by flows sharing the same links, since each link capacity is divided among all applications sharing the link.

For VBR flows, it is very difficult to evaluate the instantaneous application data rate at a specified transmission time point as the data rate fluctuates. The shortest path used in dynamic routing scenario has a better EED. Finally, with increasing packet size, EED average increases for all applications.

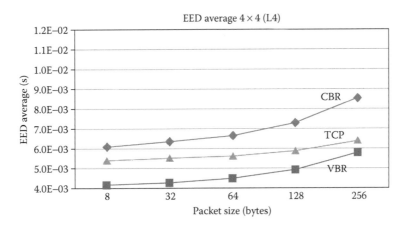

FIGURE 6.8   EED average of three applications according to packet size with router (L4).

6.6.2 Throughput

Figure 6.9 shows the variation of the concerned metric Thp average with respect to the available packet size for CBR, VBR, and TCP applications with router (L4).

We observe in Figure 6.9 that CBR traffic data rate does not fluctuate during transmission.

The wires in the links of the NoC are shared by applications, and the curve of CBR and Thp average is near smooth during packet size variation. While VBR gives the best Thp average variation, TCP gives worst results.

6.6.3 Rate of Packet Loss

Figures 6.10 through 6.12 show the variation of the concerned metric, that is, packet loss versus the available packet size for CBR, VBR, and TCP applications. Packet loss is due to the queuing discipline of switch for the proposed dynamic protocol scenario. From Figure 6.13, it is seen that TCP gives the highest rate of packet loss variation; we can also observe that the rate of packet loss variations of TCP and VBR increases with an increase in packet size and decrease in CBR.

Thus, application traffic generally tries to ensure the delivery of the packet over the transmission network channels during the time increase of all incoming flow. However, during the propagation of data, some packets have been lost due to lack of buffering capacity and priority queuing over a longer period of time.

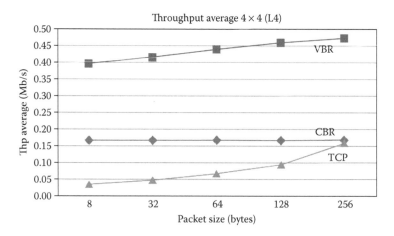

FIGURE 6.9   Average of three applications according to packet size with router (L4).

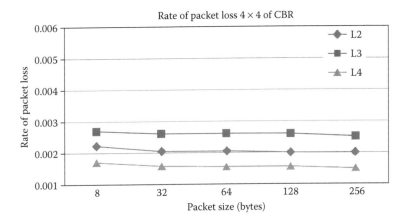

FIGURE 6.10   Rate of packet loss of CBR according to packet size.

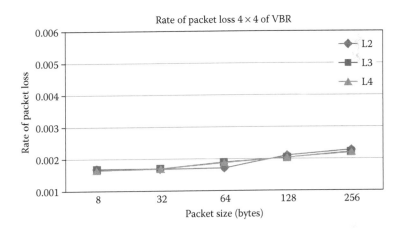

FIGURE 6.11   Rate of packet loss of VBR according to packet size.

6.6.4 QoS Measurements and Analysis

Referring to the proposed model (Equation 6.11) and Figure 6.3, QoS can be presented by the following formula:

$$QoS = QoS_0 + tr\left[diag(\beta_i)_{1\le i\le m} \times \left(\alpha_{ij}\right)_{\substack{1\le i\le m \\ 1\le j\le n}} \times \left(\hat{p}_{ij}\right)_{\substack{1\le i\le m \\ 1\le j\le n}} \right] \qquad (6.12)$$

where QoS_0 represents the minimum basic required QoS (in our case, we chose $QoS_0 = 10\%$ of the value of the ideal QoS), α_{ij} and β_i are, respectively,

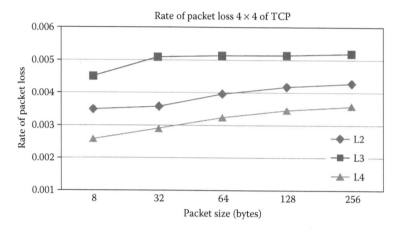

FIGURE 6.12   Rate of packet loss of TCP according to packet size.

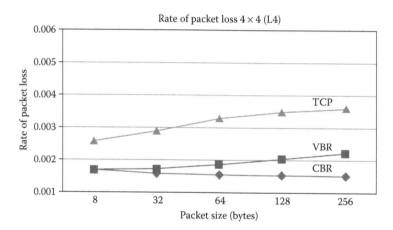

FIGURE 6.13   Rate of packet loss of three applications according to packet size with router (L4).

prioritization factors of parameters and applications, arbitrarily fixed referring to the following equation (for one application ap_i):

$$\sum_{j=1}^{n} (\alpha_{ij}) = 1 \quad \text{and} \quad \sum_{j=1}^{n} (\beta_j) = 1 \qquad (6.13)$$

We chose the parameter prioritization factor α_{ij}, the application prioritization factor β_j, and the minimum acceptable value QoS_0, as shown in Figures 6.14 and 6.15.

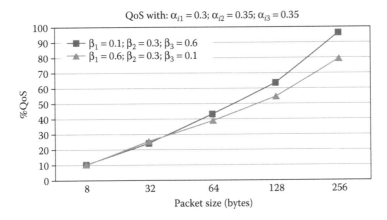

FIGURE 6.14 Percentage QoS with parameter prioritization factors ($\alpha_{i1} = 0.3$; $\alpha_{i2} = 0.35$; $\alpha_{i3} = 0.35$) of three applications according to packet size.

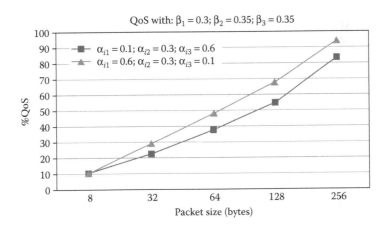

FIGURE 6.15 Percentage QoS with application prioritization factors ($\beta_1 = 0.3$; $\beta_2 = 0.35$; $\beta_3 = 0.35$) of three applications according to packet size.

We used the model (Equation 6.12) and considered three QoS performance parameters, such as PLR as p_{i1}, EED as p_{i2}, and Thp as p_{i3}, for three concurrent applications CBR ($i = 1$), VBR ($i = 2$), and TCP ($i = 3$) for different available packet sizes with router (L4).

Figures 6.14 and 6.15 show the QoS percentage in relation to the packet size, the scheduling techniques, and parameter and application prioritization factors. It appears that the percentage of QoS increases with the increase in packet size. Application prioritization factors also have an impact on the QoS values.

This network performance analysis shows that all the three applications provide QoS value for applications traffics.

The approach proposed in Nasri (2011b) is based on the deficit weighted round-robin technique for the management of the data queuing for one application service. However, it provides QoS value rate between 0% and 94% according to the application rate, whereas the overall improvement in our extended approach is the combination of many QoS parameters of many applications. Furthermore, we observe in Figure 6.14 that QoS of all application flows begin by 10% as a minimum basic required QoS value rate and has a highest value rate between 79% and 96% due to parameter prioritization factors. In contrast, in Figure 6.15, the highest value rate is between 82% and 94% due to the application prioritization factors. This reflects the overhead that comes with most QoS-enforced systems. By varying application and parameter prioritization factors in Figures 6.14 and 6.15, the model approach (Equation 6.12) gives for each packet size a quantifiable representation of QoS and helps to make up the efficiency of the QoS metrics evaluation.

Although the QoS is qualitative and not measurable, we have proposed a new approach of its quantifiable representation.

6.7 CONCLUSIONS AND PERSPECTIVES

Until now, there is no standard method of the QoS measurement. Some researchers have defined the QoS from its parameters in a defined environment.

In this work, we proposed an expended approach of QoS metrics modeling for NoC based on prioritization factors of multiparameter and multiapplication environment.

QoS parameters are an important metric in NoC performance evaluation. For this purpose, we analyzed QoS metrics based on QoS parameter specification using three fundamental measures of network performance such as EED, Thp, and PLR in NoC nodes.

These QoS parameters determine a network connection speed subject of multiple applications in a dynamic routing environment.

We have also showed that dynamic routing improves better performance and reduces congestion by choosing more optimal links.

However, by implementing QoS–IP module, the communication system became self-regulating.

It can produce stability and reduce the effect of NoC communication problems and the gap between the measurement and the required end-to-end QoS–NoC.

Furthermore, we have evaluated an extended approach of QoS metrics on a 4×4 mesh NoC, based on QoS parameters estimation, in which a minimum required QoS value (QoS_0) is integrated.

EED and Thp results have shown that VBR application is better than the CBR and TCP applications. A good PLR is offered by CBR application.

We also have shown that metrics of QoS during NoC communication processes are affected by the packet size and increase with parameter and application prioritization factors.

This work has completed with a study and development of a QoS metrics modeling for NoC on multiparameter and multiapplication environment where experimental results show the fitness of our approach.

The proposed integrated QoS management system is important in making QoS capable communication systems able to deal efficiently with the increasing variety of applications.

Therefore, further research should be done in order to find additional simplifications to and improvements in the approach.

A QoS metrics measurement based on dynamic routing simulation shaped on the behavior of different NoC topologies with multiparameter and multiconcurrent application environment will constitute future work.

REFERENCES

Al Faruque, M. A., Ebi, T., and Henkel, J. (2010). AdNoC: Runtime adaptive network-on-chip architecture. In *IEEE Transactions on Very Large Scale Integration (VLSI) Systems*, vol. 99 (pp. 257–269).

Beigne, E., Clermidy, F., Vivet, P., Clouard, A., and Renaudin, M. (2005). An asynchronous NOC architecture providing low latency service and its multi-level design framework. In *Proceedings of the 11th IEEE International Symposium on Asynchronous Circuits and Systems* (pp. 54–63).

Benini, L., and De Micheli, G. (2002). Networks on chips: A new SoC paradigm. *IEEE Computer*, 35(1), 70–78. doi:10.1109/2.976921.

Benini, L., and De Micheli, G. (Eds.). (2006). *Networks on Chips: Technology and Tools*. San Francisco, CA: Morgan Kaufmann.

Bjerregaard, T., and Mahadevan, S. (2006). A survey of research and practices of network-on-chip. *ACM Computing Surveys*, 38, 71–121. doi:10.1145/1132952.1132953.

Bolotin, E., Cidon, I., Ginosar, R., and Kolodny, A. (2004). QNoC: QoS architecture and design process for network on chip. *Journal of Systems Architecture, Special issue on Networks on Chip*, 50(2–3), 105–128. doi:10.1016/j.sysarc.2003.07.004.

Chunlin, L., and Layuan, L. (2008). Cross-layer optimization policy for QoS scheduling in computational grid. *Journal of Network and Computer Applications*, Elsevier, 31, 258–284. doi:10.1016/j.jnca.2006.12.001.

Dally, W. J., and Aoki, H. (1993). Deadlock-free adaptive routing in multicomputer networks using virtual channels. *IEEE Transactions Parallel Distributed Systems*, 4(4), 466–475. doi:10.1109/71.219761.

Dally, W. J., and Towles, B. (2001). Route packets not wires: On-chip interconnection networks. In *Proceedings of Design Automation Conference (DAC'01)* (pp. 648–689). New York: ACM, IEEE. doi:10.1109/DAC.2001.156225.

Das, R., Mutlu, O., Moscibroda, T., and Das, C. R. (2009). Application-aware prioritization mechanisms for on-chip networks. In IEEE (Ed.), *MICRO 42: Proceedings of the 42nd Annual IEEE/ACM International Symposium on Microarchitecture* (pp. 280–291). New York: ACM, IEEE.

De Micheli, G., Seiculescu, C., Murali, S., Benini, L., Angiolini, F., and Pullini, A. (2010). Networks on chips: From research to products. In *Proceedings of DAC* (pp. 300–305). New York: ACM. doi:10.1145/1837274.1837352.

Goossens, K., Dielissen, J., Gangwal, O. P., Gonzalez Pestana, S., Radulescu, A., and Rijpkema E. (2005). A design flow for application-specific networks on chip with guaranteed performance to accelerate SOC design and verification. In *Proceedings of the Design, Automation and Test in Europe Conference and Exhibition (DATE'05)* (pp. 1182–1187). IEEE. doi:10.1109/DATE.2005.11.

Goossens, K., Dielissen, J., and Radulescu, A. (2005). AEthereal network on chip: Concepts, architectures, and implementations. *IEEE Design and Test of Computers*, 22(5), 414–421. doi:10.1109/MDT.2005.99.

Helali, A., and Nasri, S. (2009). Network on chip switch scheduling approach for QoS and hardware resources adaptation. *International Journal of Computer Sciences and Engineering Systems (IJCSES)*, 3(1), 29–35.

Helali, A., Soudani, A., Bhar, S., and Nasri, S. (2006). Study of network on chip resources allocation for QoS management. *Journal of Computer Science*, 2(10), 770–774. doi:10.3844/jcssp.2006.770.774.

Helali, A., Soudani, A., Nasri, S., and Divoux, T. (2005). An approach for end-to-end QoS and network resources management. *Computer Standards & Interfaces Journal*, 28, 93–108. doi:10.1016/j.csi.2004.12.002.

Joven, J., Marongiu, A., Angiolini, F., Benini, L., and De Micheli, G. (October 24–29, 2010). Exploring programming model-driven QoS support for NoC-based platforms. In *Proceedings of the 8th IEEE/ACM/IFIP International Conference on Hardware/Software Codesign and System Synthesis (CODES/ISSS '10)* (pp. 65–74), Scottsdale, AZ.

Kim, J., Park, D., Nicopoulos, C., Vijaykrishnan, N., and Chita, D. R. (2005). Design and analysis of a NoC architecture from performance, reliability and energy perspective. In *ACM Symposium on Architecture for Networking and Communications Systems* (pp. 173–182), Princeton, NJ: IEEE. doi:10.1109/ANCS.2005.4675277.

Li, B., Zhao, L., Iyer, R., Peh, L-S., Leddige, M., Espig, M., Lee, S. E., Newell, D. (2011). CoQoS: Coordinating QoS-aware shared resources in NoC-based SoCs. *Journal of Parallel and Distributed Computing*, 71(5), 700–713. doi:10.1016/j.jpdc.2010.10.013.

Lima, S., Carvalho, P., and Freitas, V. (2004). Distributed admission control for QoS and SLS management. *Journal of Network and Systems Management, Special Issue on Distributed Management*, 12(3), 397–426. doi:10.1023/B:JONS.0000043687.34623.32.

Lima, S., Carvalho, P., and Freitas, V. (2005). Self-adaptive distributed management of QoS and SLSs in multiservice networks. In *Proceedings of the 9th IFIP/IEEE on Integrated Network Management* (pp. 411–424). IEEE. doi:10.1109/INM.2005.1440811.

Liu, M., Xu, S., and Sun, S. (2012). An agent-assisted QoS-based routing algorithm for wireless sensor networks. *Journal of Network and Computer Applications*, Elsevier, 35(1), 29–36. doi:10.1016/j.jnca.2011.03.031.

Marescaux, T., and Corporaal, H. (2007). Introducing the SuperGT network-on-chip; SuperGT QoS: More than just GT. In *Proceedings of the Design Automation Conference (DAC'07)* (pp. 116–121). San Diego, CA: ACM, IEEE.

Murali, S., Coenen, M., Radulescu, A., Goossens, K., and De Micheli, G. (2006). A methodology for mapping multiple use-cases onto networks on chips. In *Conference on Design, Automation and Test in Europe* (pp. 118–123), Munich.

Nachiappan, C., Mishra, A. K., Kandemir, M. T., Sivasubramaniam, A., Mutlu, O., and Das, C. R. (September 19–23, 2012). Application-aware prefetch prioritization in on-chip networks. In *Proceedings of the 21st International Conference on Parallel Architectures and Compilation Techniques* (pp. 441–442), Minneapolis, MN. doi:10.1145/2370816.2370886.

Nasri, S. (2011a). Network on chip: A new approach of QoS metric modelling based on calculus theory. *International Journal of Communications, Network and System Sciences*, 3(5), 53–60.

Nasri, S. (2011b). New approach of QoS metric modeling on network on chip. *International Journal of Communications, Network and System Sciences*, 4(5), 351–355. doi:10.4236/ijcns.2011.45040.

Plestys, R., Vilitis, G., Sandonavicius, D., Vaskeviciute, D., Kavaliunas, R., and Kaunas, R. (2007). The measurement of grid QoS parameters. In *IEEE, Proceedings of the ITI 29th International Conference on Information Technology Interfaces* (pp. 703–707). Cavtat: IEEE. doi:10.1109/ITI.2007.4283857.

Rijpkema, E., Goossens, K., and Radulescu, A. (2003). Trade offs in the design of a router with both guaranteed and best-effort services for networks on chip. In *Design, Automation and Test in Europe* (DATE'03) (pp. 350–355). IEEE. doi:10.1049/ip-cdt:20030830.

Rijpkema, E., Goossens, K., and Wielage, P. (2001). A router architecture for networks on silicon. In *The 2nd Workshop on Embedded Systems (PROGRESS'2001)* (pp. 181–188), Veldhoven.

Saadaoui, A., and Nasri, S. (2012). NoC: QoS metrics modelling and analysis based on dynamic routing. *International Journal of Distributed and Parallel Systems (IJDPS)*, 3(2), 43–52. doi:10.5121/ijdps.2012.3204.

Tenhunen, J. H. (Eds.). (2003). *Networks on Chip*. Hingham, MA: Kluwer Academic.

Winter, M., Prusseit, S., and Gerhard, P. F. (2010). Hierarchical routing architectures in clustered 2D-mesh networks-on-chip. In *IEEE, SoC Design Conference (ISOCC)* (pp. 388–391). Seoul: IEEE. doi:10.1109/SOCDC.2010.5682890.

Yang, L., and Huacan, H. (2007). Grid service selection using QoS model. In *Proceedings of the 3rd International Conference on Semantics, Knowledge and Grid (SKG 2007)* (pp. 576–577). IEEE. doi:10.1109/ SKG.2007.33.

Yiping, D., Zhen, L., and Watanabe, T. (November 21–24, 2010). An efficient hardware routing algorithms for NoC. In *Conference on Tencon IEEE Region*, vol. 10 (pp. 1525–1530), Fukuoka, Japan: IEEE.

Yunus, S. A. M. J., Marsono, M. N., and Ibrahim, I. (February 13–16, 2011). Modeling router hotspots on network-on-chip. In *Proceedings of the 13th International Conference IEEE, Advanced Communication Technology (ICACT)* (pp. 896–900), Seoul, Korea: IEEE.

Zhiang, W., and Junzhou, L. (2006). The measurement model of grid QoS. In *Proceedings of the 10th International Conference on Computer Supported Cooperative Work in Design* (pp. 1–6), Nanjing, China: IEEE. doi:10.1109/ CSCWD.2006.253249.

CHAPTER 7

Ant Colony Routing for Latency Reduction in 3D Networks-on-Chip

Luneque Del Rio Souza e Silva Jr., Nadia Nedjah, and Luiza de Macedo Mourelle

CONTENTS

7.1 INTRODUCTION

Network-on-chip (NoC) is a kind of communication infrastructure that allows scalability in the design of complex integrated circuits [1–2]. The evolution in transistor integration led designers to create block-based components, termed *intellectual property* (IP) cores. Such methodology made possible the development of systems-on-chip (SoC), integrated devices composed of several IP cores. Being a full computer, the SoCs must manage the transit of data between its components, such as memory blocks and processors.

The design of NoC-based projects can be divided in many intermediary steps, like the allocation of tasks [3], the mapping of IP cores [4], and the static routing [5]. Each of these steps are optimized by computer-aided tools, in general called EDAs (electronic design automation). An ideal EDA uses the specification of the application and generates a complete system implementation. In a real design, the EDA tool optimizes characteristics of both hardware and software to achieve a solution that meets the design specification. This optimization is done in an iterative way, as seen in Algorithm 7.1.

One of the main issues in communication architectures is the increase of the *latency*, the time of transmission of data between sender and receiver. Delays will occur when more than one sender tries to use a same communication link, characterizing the *congestion* of the network. The focus of this chapter is optimization of the routing step in the design of NoC-based architectures. The search for routes that avoid congestion events can improve the execution time of the application. A bio-inspired meta-heuristics, the ant colony optimization (ACO) [6], is used to search and optimize routes within the NoC infrastructure. This algorithm has proven to be effective for static routing in networks having 2D-mesh topology. This chapter is a study of the behavior of ACO-based routing algorithms applied to 3D networks having mesh, torus, and hypercube topologies.

ALGORITHM 7.1   GENERAL DESIGN IN NOC PLATFORM

Require: IP repository;
Require: Design constraints;
  1: **while** Design constraints not met **do**
  2:    Perform optimization steps;
  3: **end while**
  4: **return** System system implementation

The organization of this chapter is shown as follows. Section 7.2 presents some characteristics of NoC topologies. Relevant works on routing algorithms are introduced in Section 7.3. In Section 7.4, there is an overview of the ACO meta-heuristics. The model used to characterize the routing problem is detailed in Section 7.5. The application's specifications and its use in NoC platform are presented in Section 7.6. In Section 7.7, the results of simulation experiments are discussed. The conclusion of this work, in which some relevant future work is pointed out, is presented in Section 7.8.

7.2 DIRECT NETWORKS TOPOLOGIES

In a direct network, the nodes are composed by a switch, a network interface, and a processing element. In the context of NoCs, the processing elements that constitute the system are also called *resources* [1]. Nodes are connected together via point-to-point channels or links [7]. The information is transmitted between switches divided in smaller parts, such as messages or packet. This whole concept is an analogy to traditional computer networks systems, but with all components encapsulated in a same integrated circuit. Multiprocessor systems-on-chip (MPSoCs) [8] can be implemented using NoC as communication infrastructure, for applications that explore high levels of parallelism.

The way that interconnection is organized defines a *network topology*. Many topologies are used in early commercial parallel computers, like *mesh* in Intel Paragon [9], *torus* in the Cray T3D [10], and *hypercube* in nCUBE-3 [11]. In Figure 7.1a and b meshes and tori topologies are shown

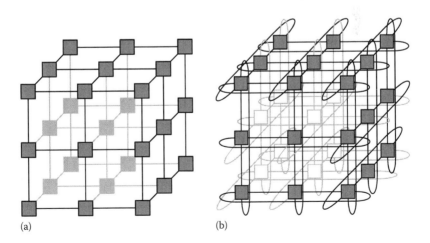

(a)   (b)

FIGURE 7.1   3D (3 × 3 × 3) mesh (a) and torus (b).

in 3D networks. In the mesh topology, switches are connected linearly in each dimension; in torus topology, switches are also connected, but forming a ring in each dimension [12].

Figure 7.2 shows the interconnection of nodes in hypercube or n-cube topology. Only two nodes can be connected in a same dimension. If more nodes are needed, then more are the number of dimensions that each switch can access.

For a same number of nodes, the choice of a particular topology impacts the trade-off between parameters associated with physical constraints of the network. These parameters include *bisection bandwidth* and *node size* [7]. The bisection bandwidth is the minimum number of links that must be selected when the network is divided into two parts with the same number of nodes. The node size is the number of links per switch. For n-dimensional mesh, torus, and hypercube, bisection bandwidth and node size are defined in Table 7.1, where n is the number of dimensions and k is the number of nodes per dimension.

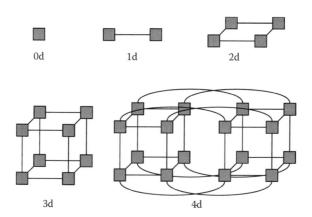

FIGURE 7.2 Hypercube with different dimensions.

TABLE 7.1   Network Characteristics in Different Topologies

Topology	Bisection Bandwidth	Node Size
nD mesh	k^{n-1}	$2n$
nD torus	$2k^{n-1}$	$2n$
nD hypercube	2^{k-1}	k

7.3 ROUTING AND RELATED WORKS

The study of efficient network routing is wide, because same techniques can be adapted to different platforms. The survey in directed networks made by Ni and McKinley [13] presents several of these techniques.

The XY algorithm is the more simple routing for networks with 2D-mesh topology. In this technique, packets are sent first in one dimension, and then in the second dimension, considering the network in a coordinated system. Due to its simplicity, the XY algorithm is widely used in computer networks and multiprocessor systems [13]. A similar routing algorithm can be used, if the topology has more than two dimensions. This is called dimension order routing (DOR). The XY algorithm was used in NoC-based systems, including the HERMES network [14] and the SoCIN network [15], and proved its efficiency due to its deadlock-free characteristic. A deadlock, or *circular wait*, occurs when a packet is waiting for a link, which is being used by another packet, which in turn is waiting for the first packet to release the link.

Although DOR achieves shortest paths, it is not able to avoid congested regions of the network, given its deterministic nature [16]. The O1Turn routing [17] is a variation of XY algorithm for 2D mesh that chooses randomly the two minimal paths between sender and receiver, the XY or the YX. The O1Turn routing achieves minimal paths, like the XY algorithm, but has better throughput because of the division of routes. The routing proposed in [18] increases the throughput by load balancing globally across the entire network. In this case, the routing is composed by two steps: first, packets are sent from source to a random intermediate node; after that, they are sent from the intermediate node to the destination. The DOR strategy is used in both steps. Both in O1Turn and in Valiant, the use of randomness adds a degree of adaptability to the routing algorithm.

Glass and Ni proposed an adaptive routing scheme that avoids both deadlocks and livelocks, called Turn Model [19]. A *turn* is defined as a change of dimension or direction of transmission of packets. In this technique, the restriction on the number of turns avoids the formation of cyclic paths in transmission, which usually yields deadlocks. The authors proposed three routing algorithms using the Turn Model: Negative-First, West-First, and North-Last. The Odd–Even Turn Model is a similar approach proposed by Chiu [20]. In this model, turns in the transmission are restricted in function of the position of switch wherein the turn can occur.

In Duato [21], the author addresses the mathematical foundations of routing algorithms, having as main interest the design of adaptive routing algorithms for multicomputer networks. Nonetheless, most of the concepts are directly applicable to NoCs. In his work, the theoretical foundations for deadlock-free adaptive routing in wormhole networks are considered.

Meta-heuristics has been used as intelligent approaches for efficient routing algorithms. This is the case of AntNet [22], a dynamic routing scheme for telecommunication networks. In [23–26], the routing problem is dealt with only in the case of the 2D-mesh topology. In contrast, this chapter treats the problem with the perspective of 3D topologies, such as torus and hypercube.

7.4 ANT COLONY OPTIMIZATION

In recent years, several meta-heuristics have been used in solving various engineering problems. Multiagent methods optimize problems by iteratively trying to improve a candidate solution with regard to a given quality metrics.

The ACO [6] is a meta-heuristics biologically inspired by the behavior of real ants. Ant algorithms have been used in the search of solutions of NP-hard combinatorial problems [27]. Real ants can successfully find shorter paths in a multiple-path environment using *indirect communication* through *pheromones*. The operation of chemical pheromones was well explained by Goss et al. [28] with the double bridge experiment. An ant colony and a food source are connected by two bridges. Ants forage the food walking in bridges and leaving the pheromones, highly volatile chemicals that can guide other ants. Those ants that find the food source return to the colony leaving more pheromone in the path. The ants that choose the shorter path can go and return in a shorter time, reinforcing the pheromone before it evaporates. After some time, due to its positive feedback, more ants will be in the shorter path, while the pheromone in the bigger bridge evaporates. The capacity of sensing different concentrations of pheromone makes ants able to find paths not only in the double bridge experiment, but also in several kinds of complex routes. This is exactly the property that inspired ant algorithms: the search for paths in graphs, that is, paths that represent the solution in combinatorial problems represented by graphs.

The main concepts of ACO can be summarized by: (*1*) the colony of *artificial ants*, a population of agents that searches by the solution of a given problem in an iterative way; (*2*) the use of *stochastic rules* for building

---

ALGORITHM 7.2   GENERAL FORM OF ACO META-HEURISTICS

```
1: start algorithm parameters and pheromone;
2: while stop condition not met do
3:    build ant solutions;
4:    perform local search (optional);
5:    update pheromone trails;
6: end while
```

---

solutions; and (3) the use of local *artificial pheromones*, the experience of previous iterations of algorithm that can guide the artificial ants in the colony in the construction of better solutions for the problem. Ant algorithms follow the general form of the ACO meta-heuristics [29], shown in Algorithm 7.2.

In a given ant-based algorithm, each solution is built by ants using only two kinds of information: the characteristics of the representative graph, which is intrinsic of each problem, and the pheromone, which is the information of the solutions built by ants in previous iterations of the algorithm. Both types of information are locally accessible. The characteristics of the problem are static; they do not change during the execution of algorithm. The pheromone, in contrast, changes in each iteration. In all steps of solution construction, each ant deposits a quantity of pheromone locally. Better solutions are associated to large deposits of pheromones, while not reinforced paths of not-so-good solutions evaporate in time. In the context of social insects, this is a typical form of indirect communication, also called *stigmergy* [30].

7.5 ACO-BASED ROUTING

As said in Section 7.4, the ACO meta-heuristics can be used in the search of paths, being well suitable in high-load routing problems. In this work, an ACO-based algorithm is used in the optimization of static routing in NoC platforms. The search is performed in graphs that represent the network topology. The ant algorithm presented in this work differs from simple ones by the use of *multiple colonies*, with each colony having its own ants and pheromones. For multiple packets sent in the network at the same time, each colony will optimize the route of each one. However, some kind of information must be shared in order to minimize the latency, thus improving the performance of the system as a whole. The load of network is updated, in all iteration cycles, using the solutions found in previous iterations. In fact, the load of the network, that is, the links used in the transmission of a packet, is a characteristic of the problem. In our algorithm, ants know two information: the pheromone concentration in the surrounding nodes, which is updated in each iteration; and the load on a node, which is related to the waiting time in each of the possible transmission directions and updated only at the end of a cycle.

7.5.1 Network Model

To enable the use of ACO in routing, it is necessary to define a network model. In this work, the switch has communication ports with resource and neighboring switches and does not use virtual channels. It uses a *Stop/Go* flow control, with the transmission being interrupted when the packet is blocked [7]. As arbitration, it uses a *first come first serve* strategy, where requests are granted based on their order of arrival. Wormhole [13] is the switching technique adopted. It works by dividing the message in *flits* (flow units) that have the width of the communication channel. These flits are sent through the network links sequentially.

$$T_{\text{packet}} = t_{\text{flit}} \cdot \left(D + \frac{L + L_{\text{delay}}}{W} \right) \qquad (7.1)$$

Equation 7.1 describes the latency in packet transmission for networks with wormhole switching, where t_{flit} is the transmission time of a packet between two switches, D is the quantity of switches in the path, L is the length, in bits, of a packet, W is the length of a channel, and L_{delay} is the number of bits not transmitted or delayed because of congestion. Another interpretation for the latency time of Equation 7.1 is the composition of three times: the time to send the first flit from source to destination; the time to send all the flits of the packet; and the waiting time, in case of congestion.

7.5.1.1 3D-ACR

The *three-dimensional ant colony routing* (3D-ACR) is an adaptation of the *elitist ant system* [6] algorithm in the search and optimization of paths in NoC static routing. In previous works, ant algorithm was effective for routing in 2D-mesh [5,31] topologies. In this work, it is shown that 3D-ACR can be used in different topologies, like torus and hypercube, with some minor changes.

Given a set of packets that must be transmitted simultaneously over the network, 3D-ACR tries to find paths that minimize the transmission latency. The algorithm works as shown in Figure 7.3. It uses g interdependent colonies. For each colony, m ants search for paths for routing a single packet using their own pheromone. The order of colonies must be followed, so that the kth ant of a colony knows the path found by all kth ants of other colonies. After all $m \times g$ ants have found paths, the pheromone is updated, and the cycle restarts.

The set of nodes chosen by ants are probabilistically selected based on Equation 7.2, wherein p_{ij}^k is the chance of ant k going from node i to node j.

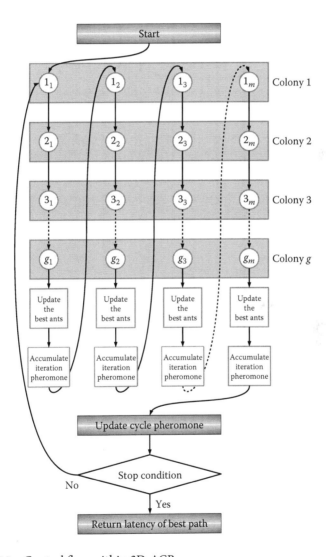

FIGURE 7.3 Control flow within 3D-ACR.

$$p_{ij}^{k}(t)=\begin{cases}\dfrac{\tau_{j}(t)^{\alpha}\cdot\eta_{ij}^{\beta}}{\sum_{k\in a_{k}}\tau_{k}(t)^{\alpha}\cdot\eta_{ik}^{\beta}}, & \text{if } j \in a_{k} \\ 0, & \text{otherwise}\end{cases} \quad (7.2)$$

The probabilistic rule depends on two quantities: τ_{j}, the pheromone concentration in a given direction, and η_{ij} that is related with the network load C_{ij}, as defined in Equation 7.3. Pheromone and load are weighted in the algorithm by two important constants, α and β

$$\eta_{ij} = \frac{1}{C_{ij}} \quad (7.3)$$

There is a step of *pheromone update* at the end of each iterative cycle, with deposit and evaporation of pheromones following Equation 7.4.

$$\tau_{t+1} = (1-\rho)\cdot\tau_{t} + \sum_{k=1}^{m}\Delta\tau^{k} + \tau_{\text{elite}} \quad (7.4)$$

The colony pheromone of the previous iteration, τ_{t}, decreases in all nodes using evaporation rate constant ρ. The pheromone is also reinforced by the contribution $\Delta\tau$ from all m ants of the colony in the current cycle. The update rule of Equation 7.4 uses the reinforcement of pheromone from ants that found the best solutions. This is the pheromone of *elitist ants*, τ_{elite}. Equation 7.5 defines the pheromone deposit of a single ant k.

$$\Delta\tau^{k} = \frac{Q}{L_{k}} \quad (7.5)$$

The ants which found longer paths, or which found paths with congestion, have high values of L. Consequently, the total amount of deposited pheromone, $\Delta\tau$, is low.

7.6 APPLICATIONS MODEL AND MAPPING

NoC-based architectures are used, in most cases, in embedded systems. In such systems, there is some kind of *application*, a piece of software that the dedicated hardware must execute. In the design of this type of systems, an EDA tool is usually responsible for tuning the characteristics of the NoC to the application, so that its execution is most efficient.

7.6.1 Task Graphs

In order to simplify, the specification of computational applications is usually represented by *task graphs* in early stages of NoC-based systems design.

This is an abstract data structure composed of fragments of the application, called tasks, and the data exchange between these tasks. The structure of tasks and communication depends on the adopted abstraction level. In general, the application task graph is an acyclic directed graph $TG = G(T, D)$, where T is a set of tasks and D is a set of edges. Tasks are specific computations, like arithmetic operations or data conversions. Edges are the data dependencies, the amount of data that must be sent between tasks.

7.6.2 Random Mapping

In Section 7.1 the concept of EDA was introduced. This kind of tool performs optimizations in different aspects of NoC to increase the efficiency in application execution. These optimizations include routing and IP allocation and mapping. While the intuit of allocation is the selection of IPs from a repository for use in the tasks execution, the mapping is the spatial organization of those IP cores in the NoC topology.

As already stated, routing is the main target in this chapter. However, to perform this optimization step, allocation and mapping steps must have been performed previously, so that the routing algorithm can use the information associated with IP cores. In this work, the task allocation is not treated. In other words, it is considered that this step was done in other work. The mapping, however, must be performed. In order to exploit communication paths, a simple *random mapping* process was chosen. Tasks from *application graph* are associated with nodes in *topology graph* in a random way.

The mapping step has the number of nodes in a given regular topology, N, as constraint. This number of nodes can be, at least, equal to the number of tasks in the application, P. These two quantities are related by Equation 7.6 for n-dimensional mesh or torus topology.

$$N = \left\lceil \sqrt[n]{P} \right\rceil^n \tag{7.6}$$

The random mapping of IPs was chosen in order to explore the performance of routing. The mapping step can use intelligent techniques [4,32] to efficiently position IP cores in places that optimize the data transmission. If a sophisticated mapping technique was used, the impact of routing optimization would be minimal. The effort of this work is to use the optimized routing for congestion avoidance in systems with not optimal mapping. Because random mapping places the IP cores in any node of the network, this can be used to evaluate the routing in different situations.

7.7 EXPERIMENTS AND RESULTS

In this work, a simulator was implemented in MATLAB platform [33], supporting networks with wormhole switching. Networks with the following topologies are investigated: 3D mesh and 3D torus. Four routing algorithms were tested in the cycle-accurate simulator: 3D-ACR, the algorithm that is being evaluated; and DOR (XYZ), O1Turn, and Valiant, three other routing schemes for 3D topologies. There is also a study of the behavior of 3D-ACR in networks with hypercube topology. The simulations were classified into two types: tests with random traffic generated following some well-known patterns and traffic of applications mapped onto the network. The performance metrics have as time unit the simulator cycle, the time to transmit a flit between two switches.

All algorithms were executed with MATLAB Version 7.7.0.471 (R008b). Simulations were carried out on computers having Intel Core i7 950 3GHz, 8Gb RAM, and Microsoft Windows 7 Home Premium operating system.

7.7.1 Traffic Patterns

This set of simulations is based on the use of synthetic load to evaluate the different routing strategies used in the networks. The size of network with 3D topology was $5 \times 5 \times 5$ (125 nodes). Increasing the number of nodes will result in an increase in the processing time of the routing algorithm. This is not really a problem, since it is an offline optimization. The parameters of the simulation are shown in Table 7.2. Each test is performed varying the routing scheme and the network load. The node that sends the packet and the receiver node is called *source destination pair*. The pairs are selected randomly, following six different distribution patterns. This use of traffic patterns is recurrent in the evaluation of communication in multiprocessor and distributed systems [7].

Three distributions, *uniform*, *hot spot*, and *local*, are called random patterns. In these cases, all nodes are chosen in a random way. In the uniform pattern, every node can be used as source or destination with the same probability. In the hot spot, all nodes can be selected as source with the same probability. However, some nodes, the hot spots, have a greater

TABLE 7.2 Parameters of Simulations with Traffic Patterns

Routing	3D-ACR, XY, O1Turn, Valiant
Patterns	Uniform, Hot spot, Local, Complement, Trans. 1 and 2
Packets	10, 20, 30, 40, 50, 60, 70, 80, 90, 100
Injection rate	10%, 20%, 30%, 40%, 50%, 60%, 70%, 80%, 90%, 100%

chance of being selected as destination. In local pattern all nodes can be selected as source, but only neighboring nodes can be selected randomly as destination nodes.

The *complement, matrix transpose 1,* and *matrix transpose 2* are called deterministic patterns. In the three patterns, the destination node is function of the position of source node, as seen in Table 7.3. Source nodes have all the same probability of being selected. The quantity *size* is the number of nodes in one dimension.

In all simulations was counted the individual latency of packets, that is, the transmission time (in simulator cycles) since the injection of the first flit by the sender node until the reception of the last flit of the same packet in the receiver node. Because several packets might be transmitted in the network at the same time, congestion events may occur, increasing the individual latency for blocked packets. The used metric is the *average latency,* that is, the sum of the individual latency of all packets divided by the number of packets. The main objective of these simulations is to verify the behavior of the average latency of packets in a given network topology using different routing algorithms under different conditions of injection rate.

The routing performance is presented in *latency/packet × injection rate* graphs. These values are the arithmetic mean of the average latency for different quantity of packets. The graphs illustrate values obtained with the different routing schemes adopted. For 3D mesh and 3D torus, the latencies for the six traffic patterns are shown, respectively, in Figures 7.4 and 7.5.

3D-ACR has obtained smaller latencies, in comparison with the other routing algorithms, in most of the cases. In 3D mesh, 3D-ACR provides the best results with respect to all the others, except for the *complement* pattern, where the latency is higher than XYZ and O1Turn. In contrast, 3D-ACR presents a different behavior in 3D torus. For this topology, 3D-ACR yields inferior quality results than the others for small injection rates (below 50%), but obtains reduced latencies at higher injection rates. In this topology, the 3D-ACR proved less susceptible to variations in the injection rate than other routing algorithms.

TABLE 7.3 Destination Nodes of Deterministic Patterns

Pattern	Source Node	Destination Node
Complement	(x, y, z)	$(size - x + 1, size - y + 1, size - z + 1)$
Transpose 1	(x, y, z)	$(size - y + 1, size - z + 1, size - x + 1)$
Transpose 2	(x, y, z)	(y, z, x)

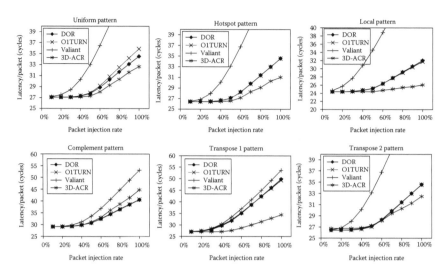

FIGURE 7.4   Latencies for six traffic patterns in 3D mesh.

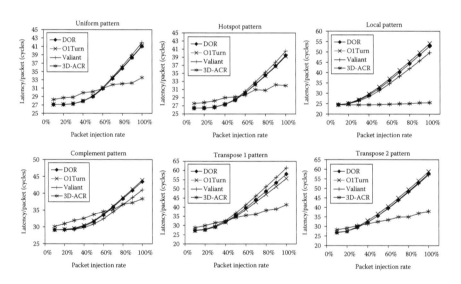

FIGURE 7.5   Latencies for six traffic patterns in 3D torus.

7.7.2 Applications Simulation

In the simulations with applications, a set of task graphs are mapped into the network with 3D topology. For each mapping, information of task graph is used in the routing, identifying which packets are transmitted at the same time. The task that generates packets and the task that depends on these packets are related to a source destination pair in the

---

ALGORITHM 7.3   MAPPING AND ROUTING OF APPLICATION

Require: Task Graph;
  1: **define** size of NoC;
  2: perform the mapping;
  3: **for** $l = 1 \rightarrow$ #*levels* **do**
  4:    get all arcs in level l;
  5:    read t_{start} of source tasks;
  6:    perform the routing;
  7:    write t_{start} of destination tasks;
  8: **end for**
  9 $t_{execution} \leftarrow t_{start}$(*last task*) + t_{comp}(*last task*)
  10: **return** routing paths, $t_{execution}$;

---

network. Several source nodes might inject packets into the network, so this is why the routes must be optimized. This process can be described by Algorithm 7.3.

7.7.2.1 Applications Repository

To evaluate the routing algorithms in 3D topologies using applications, the embedded systems synthesis benchmarks suite (E3S) [34] is used. It is a set of task graphs of real-world applications based in IP repository from the Embedded Microprocessor Consortium. The E3S was developed to use in system-level hardware researches and has characteristics of 17 embedded processors, including the execution time of 47 different tasks, the power consumption, frequency operation, and so on. The set of applications includes task graphs of common tasks in auto industry, networking, telecommunication, and office automation.

The 16 task graphs present on E3S were used in this work. The number of tasks and the number of computational levels of each task graph are shown in Table 7.4. Graphs with the same number of tasks and levels represent purely sequential applications, so the relation of these two quantities defines the application complexity.

7.7.2.2 SegImag Application

A more complex application used in the simulation is the SegImag [32]. It is a segmentation process for digital images. This application splits the image in small parts and performs the parallel computation in auxiliary processors. The complete system must contain a central processor to start the computation, auxiliary processors, and an external memory to store the images.

TABLE 7.4   Network Characteristics in Different Topologies

Label	Application Name	# Tasks	# Levels
Application 01	auto-indust-tg0	6	6
Application 02	auto-indust-tg1	4	4
Application 03	auto-indust-tg2	9	8
Application 04	auto-indust-tg3	5	5
Application 05	consumer-tg0	7	5
Application 06	consumer-tg1	5	4
Application 07	networking-tg1	4	4
Application 08	networking-tg2	4	4
Application 09	networking-tg3	4	4
Application 10	office-tg0	5	4
Application 11	telecom-tg0	4	4
Application 12	telecom-tg1	6	5
Application 13	telecom-tg2	6	5
Application 14	telecom-tg3	3	3
Application 15	telecom-tg4	3	3
Application 16	telecom-tg5	2	2

Previous implementations of SegImag [35] consider the number of auxiliary processors being parameterized. Increasing the number of processors affects the degree of parallelism, improving the execution time of segmentation process. In this work, the application is a specific implementation of SegImag, where image is divided into four parts. The task allocation is based on the work of Da Silva et al. [32], which uses processors from E3S repository. The relevant information about each task is shown in Table 7.5.

7.7.2.3 Results of Tests with Applications

In this set of tests is counted the imposed total execution time of the applications mapped on NoC, with four different routing algorithms. This execution time has two components. The first is the time for execution of a sequence of tasks in a critical path, that is, the path in the task graph with

TABLE 7.5   Tasks of E3S Used in SegImag Application

Label	Task Name	Id	Time (ns)
SI	Decompress JPEG	455	7×10^7
PI	Complex FFT	449	1.2×10^5
PF	Basic floating point	371	8.9×10^2
PCV	Autocorrelation	439	6.9×10^4
MI	Compress JPEG	454	8.7×10^7

bigger length. This task graph characteristic is the same in all executions of a given application. The second is the time necessary for communication between tasks. The *packet delay* is defined as the difference between the measured latency and the calculated latency of transmission in the network without congestion.

As done in the tests with traffic patterns, 3D-ACR, XYZ, O1Turn, and Valiant routing algorithms were used in simulations with 3D topologies. The obtained packet delays in 3D mesh and 3D torus are shown in Figures 7.6 and 7.7, respectively. The presented numeric figures are the means of the packet delay for 10 different mappings.

With respect to Valiant, Figure 7.6 shows that some delays are higher than others. This occurred because Valiant does not guarantee minimal routing. In applications represented by *Application* 03, *Application* 05, and *Application* 07, the congestion affects the packet delay. For these applications, the algorithm that optimizes routes—3D-ACR—obtained smaller latencies than the deterministic ones. Results of packet delays in the SegImag application are shown in Figure 7.8. Again, 3D-ACR obtained smaller latencies than the other considered algorithms.

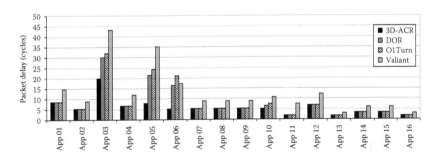

FIGURE 7.6   Packet delay of applications in 3D mesh network.

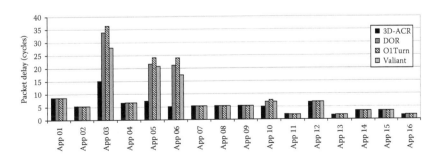

FIGURE 7.7   Packet delay of applications in 3D torus network.

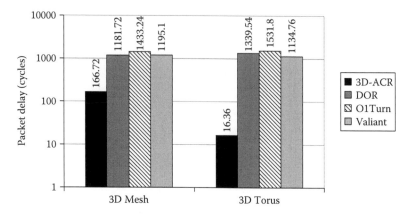

FIGURE 7.8   Packet delays in the SegImag application.

7.7.3 Simulations Using Hypercube Topology

The hypercube differs from other topologies because each dimension has only two nodes. Due to this fact, the tests conducted with the other topologies could not be used. Nevertheless, the interest is to check 3D-ACR with the use of this type of topology. In this analysis, 3D-ACR was compared only with DOR.

A network with 6-dimension hypercube topology, that is, 64 nodes, was subjected to a random traffic following the *uniform* pattern.

The *latencies/packet × injection rate* graph is shown in Figure 7.9. As it was the case for 3D-mesh topology, 3D-ACR yields latencies greater than

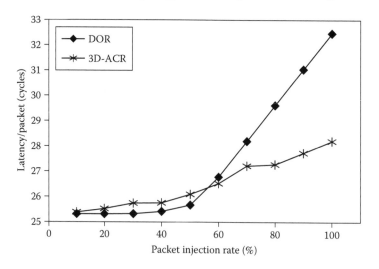

FIGURE 7.9   Latency of uniform pattern in hypercube.

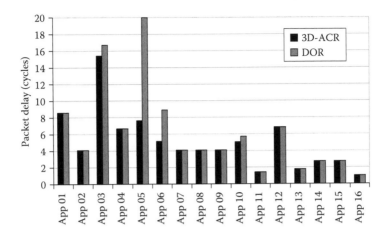

FIGURE 7.10   Packet delay of E3S applications in hypercube.

DOR for injection rates inferior than 50%. Latencies smaller than that obtained by DOR only happened at high injection rates.

In tests with applications of E3S, values of packet delay obtained by 3D-ACR were in general smaller than those obtained by DOR. These results are shown in Figure 7.10. The same behavior occurred in the case of the SegImag application, where 3D-ACR found mean delays of only 14, 86 *cycles*, against 900, 66 *cycles* for DOR.

7.8 CONCLUSIONS

This work presented 3D-ACR, an algorithm based on elitist ant system, used for static routing in NoC organized according to different topologies. A great number of simulation tests were performed, using randomly generated traffic and application task graphs. In most cases, 3D-ACR obtained superior results in comparison to deterministic and minimal routing algorithms, like DOR.

Even though satisfactory results have been obtained in the step of static routing, we think there is still room for improvements. One such improvement concerns the algorithm scalability: search time in 3D topologies is much larger than in 2D topologies. To restructure the algorithm in parallel is a possibility. In future works, in addition to changes in ACO-based routing, it would be interesting to build hybrid meta-heuristics, for example, a combination with genetic algorithms.

REFERENCES

1. L. Benini and G. De Micheli. Networks on chips: A new SoC paradigm. *Computer*, 35(1):70–78, 2002.

2. O. Ozturk and D. Demirbas. Heterogeneous network-on-chip design through evolutionary computing. *Int. J. Electr.*, 97(10):1139–1161, 2010.

3. M.V.C. Da Silva, N. Nedjah, and L.M. Mourelle. Optimal IP assignment for efficient NoC-based system implementation using NSGA-II and MicroGA. *IJCIS*, 2(2):115–123, 2009.

4. N. Nedjah, M.V.C. Da Silva, and L. M. Mourelle. Customized computer-aided application mapping on NoC infrastructure using multi-objective optimization. *J. Syst. Arch Euromicro J.*, 57(1):79–94, 2011.

5. L.S. Junior, N. Nedjah, and L. de Macedo Mourelle. Static packet routing in noc platform using aco-based algorithms. In *Intelligent Data Engineering and Automated Learning-IDEAL*, pages 84–91. Springer, 2012.

6. M. Dorigo, V. Maniezzo, and A. Colorni. Ant system: Optimization by a colony of cooperating agents. *IEEE Trans. Syst. Man Cybern., Part B Cybern.*, 26(1):29–41, 1996.

7. J. Duato, S. Yalamanchili, and L. Ni. *Interconnection Networks: An Engineering Approach*. Morgan Kaufmann, 2003.

8. L.M. Mourelle, R.E. Ferreira, and N. Nedjah. Migration selection of strategies for parallel genetic algorithms: Implementation on networks on chips. *Int. J. Electr.*, 97(10):1227–1240, 2010.

9. R. Esser and R. Knecht. Intel paragon xp/s-architecture and software environment. In *Anwendungen, Architekturen, Trends, Seminar*, pages 121–141. Springer-Verlag, Berlin, Heidelberg, 1993.

10. R.E. Kessler and J.L. Schwarzmeier. Cray t3d: A new dimension for cray research. In *Compcon Spring'93*, Digest of Papers., pages 176–182. IEEE, 1993.

11. B. Duzett and R. Buck. An overview of the nCUBE 3 supercomputer. In *Frontiers of Massively Parallel Computation*, 1992., Fourth Symposium on the, pages 458–464. IEEE, 1992.

12. J.L. Hennessy and D.A. Patterson. *Computer Architecture: A Quantitative Approach*. Elsevier, 2012.

13. L.M. Ni and P.K. McKinley. A survey of wormhole routing techniques in direct networks. *Computer*, 26(2):62–76, 1993.

14. F. Moraes, N. Calazans, A. Mello, L. Moller, and L. Ost. Hermes: An infrastructure for low area overhead packet-switching networks on chip. *Integr VLSI J.*, 38(1):69–93, 2004.

15. C.A. Zeferino and A.A. Susin. Socin: A parametric and scalable network-on-chip. In *Proceedings of the 16th Symposium on Integrated Circuits and Systems Design. SBCCI*, pages 169–174. IEEE Computer Society, Washington, DC, 2003.

16. H. Sullivan and T.R. Bashkow. A large scale, homogeneous, fully distributed parallel machine., In *Proceedings of the 4th Annual Symposium on Computer architecture. ISCA*, pages 105–117, New York, 1977.

17. D. Seo, A. Ali, W-T. Lim, N. Raque, and M. Thottethodi. Near-optimal worst-case throughput routing for two-dimensional mesh networks. In *Proceedings of the 32nd Annual International Symposium on Computer Architecture, ISCA '05*, pages 432–443. IEEE Computer Society, Washington, DC, 2005.

18. L.G. Valiant and G.J. Brebner. Universal schemes for parallel communication. In *Proceedings of the Thirteenth Annual ACM Symposium on Theory of Computing, STOC*, pages 263–277, New York, 1981.

19. C.J. Glass and L.M. Ni. The turn model for adaptive routing. In *SIGARCH Comput Arch News*, 20:278–287. ACM, 1992.

20. G.M. Chiu. The odd-even turn model for adaptive routing. *IEEE Trans. Parallel Dist. Syst.*, 11(7):729–738, 2000.

21. J. Duato. A new theory of deadlock-free adaptive routing in wormhole networks. *IEEE Trans. Parallel Dist. Syst.*, 4(12):1320–1331, 1993.

22. G. Di Caro and M. Dorigo. Antnet: Distributed stigmergetic control for communications networks. *J. Art. Int. Res.*, 9:317–365, 1998.

23. L.S. Junior, N. Nedjah, and L. de Macedo Mourelle. Routing for applications in NoC using ACO-based algorithms. *Appl. Soft Comput.*, 13(5):2224–2231, 2013.

24. N. Nedjah, L.S. Junior, and L. de Macedo Mourelle. Congestion-aware ant colony based routing algorithms for efficient application execution on network-on-chip platform. *Exp. Syst. Appl.*, 40(16):6661–6673, 2013.

25. X. Zhao and M. Gu. A novel energy-aware multi-task dynamic mapping heuristic of NoC-based MPSOCs. *Int. J. Electr.*, 100(5):603–615, 2013.

26. M.V.C. da Silva, N. Nedjah, and L.M. Mourelle. Power-aware multi-objective evolutionary optimisation for application mapping on network-on-chip platforms. *Int. J. Electr.*, 97(10):1163–1179, 2010.

27. E. Bonabeau, M. Dorigo, and G. Theraulaz. *Swarm Intelligence: From Natural to Artificial Systems*. Oxford University Press, 1999.

28. S. Goss, S. Aron, J. Deneubourg, and J. Pasteels. Self-organized shortcuts in the Argentine ant. *Naturwissenschaften*, 76:579–581, 1989. doi:10.1007/BF00462870.

29. M. Dorigo, M. Birattari, and T. Stutzle. Ant colony optimization. *IEEE Comput. Intell. Mag.*, 1(4):28–39, 2006.

30. P. Grassé. La reconstruction du nid et les coordinations interindividuelles chez bellicositermes natalensis et cubitermes sp. la thorie de la stigmergie: Essai d'interprtation du comportement des termites constructeurs. *Insectes Sociaux*, 6:41–80, 1959.

31. L.S. Junior, N. Nedjah, L. de Macedo Mourelle, and F.G. Pessanha. ACO-based static routing for network-on-chips. In *Computational Science and Its Applications, ICCSA*, pages 113–124, 2012.

32. M.V.C. Da Silva, N. Nedjah, and L.M. Mourelle. Efficient mapping of an image processing application for a network-on-chip based implementation. *Int. J. High Perform. Syst. Arch.*, 2(1):46–57, 2009.

33. MathWorks. *MATLAB Version 7.7.0 (R2008b)*. The MathWorks Inc., Natick, MA, 2008.

34. R. Dick. Embedded system synthesis benchmarks suites (E3S). http://ziyang.eecs.umich.edu/ dickrp/e3s/ [Accessed May 2, 2012].
35. C. Marcon, N. Calazans, E. Moreno, F. Moraes, F. Hessel, and A. Susin. Cafes: A framework for intrachip application modeling and communication architecture design. *J. Paral. and Dist. Comput.*, 71(5):714–728, 2011.

III

Systems Codesign

CHAPTER 8

Codem

Software/Hardware Codesign for Embedded Multicore Systems Supporting Hardware Services

Chao Wang, Xi Li, Xuehai Zhou,
Nadia Nedjah, and Aili Wang

CONTENTS

8.1 INTRODUCTION

During the past few decades, heterogeneous embedded processors and reconfigurable architecture have been integrated within field-programmable gate arrays (FPGAs) to form powerful embedded systems. As we expand the use of computing devices, capabilities beyond raw performance such as flexibility, scalability, and programmability are becoming increasingly important. In order to achieve the multiobjective optimizations, FPGA-based reconfigurable multiprocessor system-on-chip (MPSoC) has been considered as one of the promising research platforms for future microprocessors design (Borkar and Chien 2011). It is widely accepted that the benefit of reconfigurable MPSoC is to regard each intellectual property (IP) core as an accelerating engine for a specific task; therefore, it can achieve satisfying performance with creditable flexibility for diverse application fields (Singh 2011). Up to now, FPGA has been one of the major venues for hardware and software codesign, such as the optimization solutions proposed in Nedjah and De Macedo Mourelle (2009), Silva (2010), and Mourelle (2010).

However, although enjoying the benefits brought by reconfigurable computing technologies, current FPGA research community is still suffering from both high design complexity and limited programmability. Due to the large gap between diverse embedded applications, a high-level abstraction for programming model as well as IP accelerator integration is becoming a key problem to be attacked. In order to minimize the redesign complexity and also to maintain the flexibility among diverse IP accelerators, service-oriented architecture (SOA) concepts can be applied as a high-level modeling approach for reconfigurable coprocessor architectures in FPGA platform. SOA concepts are originally raised and widely used in software engineering and web service research areas. It is common knowledge that the most significant advantage of SOA is the highly adoptable modules across different ranges of computing resources. Therefore from the exploration of SOA concepts' benefits, we can conclude that there are two significant advantages introducing SOA to reconfigurable MPSoC platform: First, coprocessor integration interfaces are well defined, which could help researchers to construct high-level models by adding/removing modularized processors or IP cores expeditiously. Second, given uniform high-level application programming interfaces (API), programmers are no longer able to obtain full knowledge of the hardware implementation or the scheduling scheme, which will be automatically handled by the middleware. Both features will significantly ease the burden of programmers, and reduce design complexity to build an application-specific MPSoC.

Moreover, how to evaluate the performance of the service-based multicore architectures is still a challenging issue. Of the state-of-the-art approaches, profiling technique is one of the most effective methodologies to analyze frequencies of the programs. The primary aim of profiling techniques is to record the hot spot information, which could potentially guide the performance optimization solution. There are already various mature profiler methods and creditable tools in the commercial and academic research community. For example, for source-code level tools, gprof/gcov is most widely used with C compilers to generate statistical information, including the execution time and frequencies at function and code block levels. Along with source-code level, profiling technique is also applied at assembly-code level. State-of-the-art assembly-code level profiling tools like SpixTool (Cmelik 1993) are capable of analyzing instruction behaviors with simulator-based approaches. Compared to source-level profiling, more accurate statistical information can be obtained at this level, but the huge number of instructions leads to inevitable overheads. Therefore, current profiling techniques still lack integrated solutions with effective performance tuning solutions.

To address the above challenges, this chapter proposes a novel hot spot based profiling with state-of-the-art reconfigurable optimization software/hardware codesign flow based on the advanced RISC machines (ARM) and Xilinx toolchains, named Codem. The profiling design flow utilizes a cross-compiler, a simulator, and a profiler, while the reconfigurable tools manipulate the software/hardware codesign optimization flow. This chapter also realizes a service-oriented reconfigurable architecture on FPGA prototype. In particular, we claim following contributions on soft computing:

1. Service-oriented reconfigurable hierarchical model: this chapter realizes a novel hierarchical SOA model for coprocessor modeling on FPGA platform. SOA concepts bring structural programming and well-defined servant integration interfaces, which helps researchers to construct MPSoC prototype and high-level programming paradigms for diverse applications.

2. Adaptive mapping under reconfigurable condition: this chapter presents an adaptive mapping method based on dynamic partial reconfiguration model. When hardware reconfiguration is done, tasks can be automatically distributed to IP cores for parallel execution.

3. A novel hot spot based profiling design flow with state-of-the-art reconfigurable optimization is proposed. The profiling design flow utilizes a cross-compiler, a simulator, and a profiler, while the reconfigurable tools manipulate software/hardware codesign optimization flow. We measure the demonstrative toolchains and using the classic Sort and JPEG programs. Different task scales and working modes have been applied to identify the hot spot functions as well as to evaluate the leverage between optimization levels. Experimental results including speedups for critical test cases, hardware area, and power consumptions demonstrate the performance and cost of Codem.

Before Codem architecture is introduced, we inherit the following definitions of services in our previous work in Wang et al. (2011) at first.

Service: A *service* is defined as a specific kind of function with programming interface. All services are packaged into libraries and can be invoked by function calls.

Servants: *Servants* refer to functional modules dedicated to provide services abstracted from specific tasks and implemented in hardware. All servants are IP cores packaged in structural services definition manners.

Tasks: Throughout this chapter, we use the term *tasks* to represent pure functional instances without memory access, such as an inverse discrete cosine transform (IDCT) and advanced encrypt standard (AES) tasks running on specific hardware IP accelerators. Note that the granularity of task defined in this chapter is different from general task definitions with software threads. During task execution procedure, control information (e.g., task ID and target servant) and operands are transferred through first in first out (FIFO)-based peer-to-peer links between scheduler and servants.

The remainder of this chapter is organized as below: in Section 8.2 we outline the related work including the middleware support and software techniques based on profiling. Then Section 8.3 presents the Codem architecture and principles in detail. We propose the software and hardware design flow in Section 8.4. Afterward Section 8.5 describes the experiments and results analysis for software profiling. Finally, we conclude the chapter in Section 8.6.

8.2 RELATED WORK

Reconfigurable technologies have been concentrated in the past decades (Compton and Hauck 2002). The integration of a reconfigurable fabric into a microprocessor has been extensively studied in the context of speeding up the main computation. For example, Chimaera (Hauck et al. 2004), PipeRench (Goldstein et al. 2000), Garp (Hauser and Wawrzynek 1997), and OneChip (Wittig and Chow 1995) were early projects that developed reconfigurable fabrics for specialized streaming pipelined computations. Moreover, with the rapid development of semiconductor technologies, MPSoC hardware platform has integrated abundant computing resources. Significant expansion of reconfigurable technology is pushing MPSoC design methods toward heterogeneous and customized organizing ways, such as RAMPSoC (Gohringer et al. 2008), ReMAP (Watkins and Albonesi 2010), RAMP (Wawrzynek et al. 2007), Microsoft Accelerator (Tarditi et al. 2006), and MOLEN (Kuzmanov et al. 2004). However, most of these studies aim at application-specific hardware design, which means programmers need to acquire full knowledge of the system specification and implementation to handle the tasks mapping, scheduling, and distribution manually. Therefore, the automatic parallelization degrees are still worth pursuing.

Along with the prototype platforms targeting specific hardware, there are some creditable reconfigurable hardware-flexible infrastructures: ReconOS (Lubbers and Platzner 2009), for example, demonstrates hardware/software multithreading methodology on a host OS running on the PowerPC core of modern FPGA platforms. Hthreads (Peck et al. 2006), RecoBus (Koch et al. 2009), and (Rupnow et al. 2011) are also the state-of-the-art FPGA-based reconfigurable platforms. However, of these conducted researches, the high-level programming to circuit problems have not been completely figured out, which means users need to acquire full knowledge of the system to handle the tasks distribution manually. Therefore, it is still regarded as one of the most challenging issues in the next decades, and many researchers are focusing on the reconfigurable toolchains to efficiently help researchers for performance analysis and optimization (Neely et al. 2010). So far this problem has not been completely figured out.

In contrast, the traditional software SOA concept is shifting toward architecture level, such as factored operating system (Wentzlaff and Agarwal 2009) and services-oriented multi processors (SOMP) architecture (Wang et al. 2011). The advantage of SOA is to integrate and efficiently manage various computing resources as well as to provide structural programming interfaces in different computing areas. Thus service-based approaches can provide better flexibility and extensibility at lower cost through reusable software modules. SOMP (Wang et al. 2011) gives illustrative SOA concepts to multiprocessor design, but how the framework manages the uniform IP package and high-level programming interfaces is not clearly presented. To address the weakness of current architectures, the motivation of this chapter is to introduce SOA concepts to reconfigurable MPSoC architecture design; we can get the following abstractions: (1) First, each task is regarded as a specific macro instruction. By that means, user application consisting of multiple tasks can be abstracted to a macro instruction sequence. (2) Second, each processor or IP core is abstracted to a dedicated function unit to run an abstract instruction (or in this chapter, defined as a *servant*).

The summary of the related work is presented in Table 8.1. In spite of the fact that there are numerous directions of profiling and reconfigurable optimization techniques, the integration of these two technologies has not been conveniently proposed, especially for cross-compilation-based toolchains.

TABLE 8.1 Summary of State-of-the-Art

Major Type	Reference	Features	Drawbacks
Reconfigurable accelerators	Chimaera PipeRench OneChip	Treat each IP core as an acceleration engine	Stand-alone unit, basically a coprocessor, no multicore
Reconfigurable middleware support	RAMPSoC ReMAP RAMP Accelerator	Heterogeneous multicore platform support hybrid reconfigurations	Needs special hardware to support reconfiguration
OS-level architecture	ReconOS Hthreads RecoBus	Thread-level architecture for higher level abstraction	Needs additional management overheads and support

To address the above problem, in this chapter we present a novel integrated design flow with the support from ARM and Xilinx cross-compilation toolchains.

8.3 ARCHITECTURE AND CONCEPTS

In this section, we will first introduce the proposed architecture model with hierarchical multiple layers. Thereafter, we discriminate and comment on the integrated services. Subsequently, we detail the algorithm to solve the reconfigurable mapping and communication.

8.3.1 Introducing SOA to Hardware FPGA Design

Figure 8.1 illustrates the high-level abstract diagram of SOA-based MPSoC architecture, which is composed of one scheduler processor and multiple servants. The scheduler kernel is employed for adaptive task mapping and scheduling under reconfigurable condition. Selected tasks are distributed to certain servants via structural communication interfaces. Furthermore, the scheduler kernel also provides uniform API and runtime libraries to programmers. To make a better comprehension of the specific functional operations of the scheduler, an SOA hierarchical model with three layers is constructed inside the scheduler processor.

Application layer: High-level API and SOA runtime libraries are provided to programmers in this layer. Taking advantage from SOA concepts,

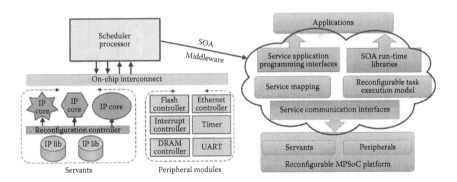

FIGURE 8.1 Architecture framework for Codem SOA-based architecture and hierarchical layer models. The left part illustrates how the scheduler processor is connected to the servants and peripheral modules. The right part is how SOA middleware is constructed inside the scheduler processor. Three hierarchical layers are abstracted between applications and reconfigurable MPSoC hardware platform.

API is separately designed from the HW/SW architecture design. Given a fixed set of API, the hardware reconfiguration and online scheduling schemes are *invisible* to the programmers.

Reconfigurable mapping layer: A reconfigurable task execution model is proposed to enable servant dynamic reconfiguration. In order to support automatic parallelization, an automatic task mapping method is employed to supervise how a task is destined to the target servant.

Communication layer: For demonstration, scheduler kernel shall spawn tasks to each servant via a pair of FIFO-based channels when the input parameters are ready. Alternatively, results are returned through interrupts in the end of task execution.

In particular, the main responsibilities of the above three layers are described in the following sections respectively.

8.3.2 Programming Model for Consistent User Behaviors

In order to maintain the consistent user programming behaviors, application layer provides structural high-level abstract programming interfaces. Both blocking and nonblocking primitives are supported: After a service is spawned, blocking interfaces shall stall the main execution thread to wait until results are returned, while nonblocking interfaces can continue the subsequent tasks. When the results are sent back via FIFO channels, an interrupt signal should stall the main execution to the interrupt handling procedure.

Figure 8.2 presents an example of annotated codes in the programming model that is extended from our previous work FPM (Wang et al. 2012) and MP-Tomasulo (Wang et al. 2013). The top part *Codemlib.h* presents an example of API implemented in the Codem runtime libraries. Both the size and I/O directions for those parameters are defined in the header file. What is required for the programmer is to include the header file in the main program, as in the lower part of the figure. When a certain API is called, a service decoding process should be operated to identify the target service, which will be forwarded to mapping layer directly.

8.3.3 Reconfigurable Mapping Scheme

In the mapping layer, services are scheduled to a destined servant. Hereby an adaptive service mapping strategy is proposed to do this job automatically.

Due to intertask dependencies, scheduler module should decide when the task can be issued, and which target servant is selected. For instance, if multiple servants are available, then the task mapping scheme should decide which servant is the best choice for current service. In order to

```
/*-- # Codemlib.h -Codem Lib Description -- */
#pragma input (idct in) output (idct_out)
void do_T_idct(int idct_out[N], idct_in[N]);

#pragma input (aes_in1, aes_in2) output (aes_out)
void do_T_aes(int aes_out[M], aes_in1[M], aes_in2[M]);

/*--Main program on scheduler processor--*/
    #include <Codemlib.h>

    main( ){
        ......
        int idct_out[N], idct _in[N];
        int aes _out[N], aes _in1[N]; aes _in2[N];
        ......
        do_T_idct(idct_out, idct _in);
        do_T_aes_enc(aes_out,aes_in1, aes_in2);
        do_T_aes_dec(idct_in,idct_out, aes_in2);
        ......
    }
```

FIGURE 8.2 Example of annotated codes in the programming model. The top part is the Codem library description including two example services. The bottom part refers to how the main program is written to call the service model.

manage servants efficiently, we employ a queue module for each type of servant, and a global servant status table. The servant status table has multiple entries, each of which contains following attributions: Servant ID is the ID of target servant. If there are multiple servants in one class, then each of them has a unique ID. Busy status indicates whether the target servant is in busy state. Service in the queue counts the number of services waiting in the queue. Moreover, considering the reconfigurable feature of Codem, the servant status table should be updated simultaneously when a hardware reconfiguration process is ready. Moreover, a hardware servant can be removed from the FPGA device when the servant is NOT in busy state and there are no queued services waiting for the servant. The queue module could only serve the pending services in FIFO order. When a service arrives and no servant is free, the service should be pushed into the queue immediately, waiting for all the prior services to finish.

After the target servant is prepared and all the input parameters are ready, the service is distributed to the target servant. When the results are returned from the interrupts, following operations are performed inside the interrupt procedure:

1. Results are stored to the destination operands.

2. Check whether the queue is empty. If not, pop the head service of the queue to communication layer. Otherwise, all the tasks are already finished.

3. Update the servant status table and spawn the service.

8.4 PROFILING-BASED DESIGN FLOW FOR EMBEDDED MULTICORE SYSTEMS

Based on the above framework, this chapter introduces hot spot–based software profiling and optimization techniques. We first describe the profiling-based design flow in this section, as illustrated in Figure 8.3.

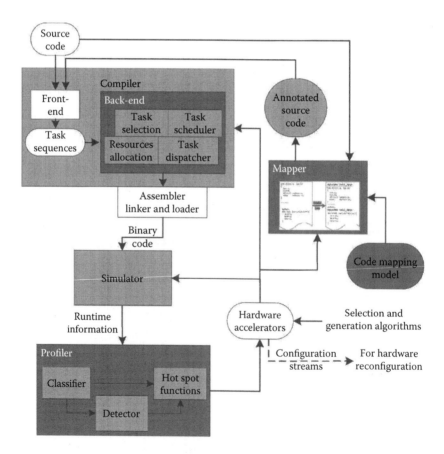

FIGURE 8.3 Example design flow with kernel profiling technologies.

8.4.1 Profiling-Based Design Flow for Embedded Systems

Generally, the integrated design flow proposed in this chapter is divided into four major phases.

8.4.1.1 Software Compilation

First of all, the source codes are compiled and linked into an executable binary file, produced by the RVDS cross toolchains. During this step, the programmers need to configure the cross toolchains, the optimization levels, and ARM/Thumb working modes, which in case could have a potential effect on the image size and performance. The annotated sources code will be generated into the task sequences, under the help of a front-end translator. Afterward, the back-end compiler will handle the task selection, resource application, and task scheduling as well as task assignment. Finally, the executable binary code is achieved.

8.4.1.2 Simulation and Online Profiling

The executable binary file is loaded by the simulator where the program is launched and executed, while the profiler starts collecting runtime information simultaneously. During the execution, the hot spot will be stored into a temporary analysis buffer. In the end of the execution, hot functions ordered by execution time and frequencies should be achieved. Furthermore, in order to inspect the detailed execution record, a trace buffer will be filled with the running instructions at every clock cycle.

8.4.1.3 Evaluation and Performance Analysis

At the end of the execution, the simulator gives the statistical analysis of the execution results to programmers for further evaluation and optimization. First, hot spot functions are the primary clues for task-level acceleration, indicating which part consumes most time and frequency. Second, for fine-grained parallelism analysis, for example, basic block level, a recorded trace file includes the following components: Instruction ID, Clock Cycle, Instruction Type, Symbolic, Address, and Opcode.

8.4.1.4 Optimization

Based on the achieved hot spot information, programmers could put their effort to the performances optimization solutions, using either software or hardware. For software and compilation level, programmers could reorganize or rewrite the source codes manually. For example, loop unrolling and parameters renaming are two state-of-the-art technologies widely applied in embedded systems. Otherwise, it is also applicable to implement hot spot functions as a coprocessor with register-transfer-level (RTL) hardware description languages. The implemented hardware could be utilized as an external stand-alone logic element, a coprocessor, or an internal function unit, due to different task granularities. State-of-the-art dynamic reconfigurable computing techniques could efficiently help researchers to add/remove the function unit implicitly without user interaction. In this stage, the hardware accelerators will be integrated and the mapper is responsible to guide the programmers for annotated source codes.

8.4.2 Reconfigurable Execution Flow

Based on the statistical results obtained from profiling techniques, an optimization method can be accessed via reconfigurable tool chain using hardware accelerators. In this chapter we present a reconfigurable tuning design flow based on Xilinx toolchains for demonstration, as is illustrated in Figure 8.4. The entire flow consists of five stages, which are marked with the related labels in the figure.

8.4.2.1 Hardware Design

Hardware description and design is the first phase in which hardware description language (HDL) (Verilog or VHDL) sources are implemented. The selection of the target function to be implemented as hardware is based on the profiling results. The RTL description code will undergo a procedure including both compilation and synthesis steps. Sample applications are designed to verify the behavior and timing correctness. Taking Xilinx tools as a demonstration, both front- and back-end simulations are operated within Xilinx ISim simulation environment, which ensures the functionality and timing correctness.

8.4.2.2 Reconfiguration Module Design

After the hardware implementations are verified, the primary design goal for each IP accelerator is met. Then we can wrap the IP for dynamic partial reconfiguration design. First both static modules and reconfigurable modules are implemented separately. The static modules refer to the fixed components in the platform, including the microprocessors, memory blocks, buses, and peripherals, while the reconfigurable modules stand for the wrapped IP cores. After all the modules are implemented, a merge operation is employed to combine both the modules together and generate the hardware platform for task execution. Design constraints from individual FPGA

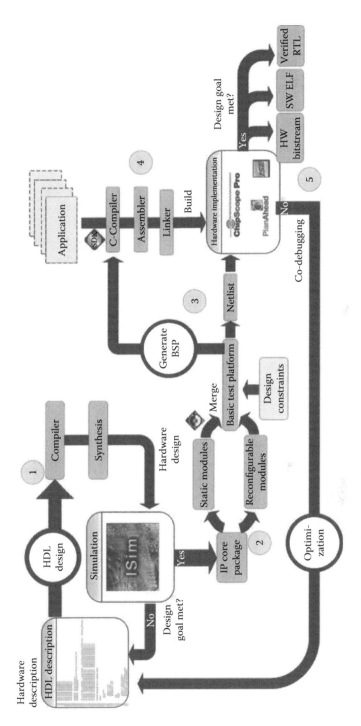

FIGURE 8.4   Example design flow based on Xilinx tools.

devices are considered during the merge process. In this chapter, we carry the module integration with Xilinx Platform Studio and IMPACT tools.

8.4.2.3 BSP and Netlists Generation

Now hardware platform design files are generated for verification and debugging, including both netlists and board support packages (BSP). Then the netlists are used for hardware bitstream generation, while the BSP files can provide fundamental hardware description running fundamental environments for diverse applications. Therefore, the BSP files are also regarded as the necessary input factors as the C compilers.

8.4.2.4 Software Compilation

Applications are built into executable files with the cross-compilation toolchains, including a C compiler, assembler, and linker. Programmers need to configure the cross toolchains, including the target ISA from the BSP file, optimization levels, and running modes, which in case could have a potential effect on the performances. Besides, hardware bitstreams in stage 3 should be programmed into FPGA chip at first, and then executable elf files can be downloaded for hardware/software co-debugging. Note that both full bitstream and partial bitstream are prepared. Static full bitstream is programmed at start-up, and partial bitstreams are used for dynamic reconfiguration at runtime.

8.4.2.5 Co-Debugging and Analysis

The co-debugging operations are constructed with Xilinx ChipScope, PlanAhead, and ISE tools. If the design goal is met, we can get the final design files including platform hardware specification bitstreams, executable elf files, and verified RTL implementations for implemented functionalities. Otherwise the performance optimization and tuning operations are considered, which leads to the hardware redesign to start over the optimization flow.

8.5 EXPERIMENTS FOR SOFTWARE PROFILING

In order to evaluate the profiling-proposed Codem design flow, we have built an experimental platform using the state-of-the-art ARM RVDS tools. For demonstration, we ported three traditional sort applications: quick, shell, and insertion sorts. The process of the applications is illustrated in Figure 8.5.

In the beginning of the application, all input digits for the three sorts are generated at random after initialization. Due to time and space constraints, the task scale indicating input number is limited to 5,000 in our experiments.

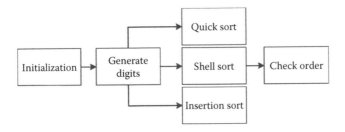

FIGURE 8.5 Processes for the sort applications.

In order to ensure that the final results are correct, after the digits are sorted, a check order function will be invoked automatically to examine the results.

The configuration for ARM RVDS toolchains is listed in Table 8.2. Besides the software configurations, ARM7TDMI simulators running on ARM/Thumb modes with –O2 optimization level are configured. Additionally, ARM tools (ARMCC, ARMASM, and ARMLNK) are employed for cross-compilation.

8.5.1 Execution Time

From the exploration of the three sort applications, the time complexities of shell, quick, and insertion sort are O(nlgn), O(nlgn), and O(n^2), respectively. We first measure the default execution time, and then obtain the running time with profiling techniques.

8.5.1.1 Default Execution Time

Figure 8.6a and 8.6c illustrates the default running time of the sort applications. The x-axis represents the task scale, ranging from 100 to 5,000, while the y-axis stands for the execution time counted in clock ticks. It is common knowledge that the insertion sort takes more time than shell and quick sort, while the shell sort takes 10% ~ 23% extra time more than quick sort, depending on the randomly generated digits. Growing with the task

TABLE 8.2   Experimental Platform Configurations

Software	Test Cases	Input	Check Order
	Shell/Quick/Insertion	Random	Automatic
Hardware	**Architecture**	**Mode**	**Optimization**
	ARM7TDMI	ARM/Thumb	–O2
Tools	**Compiler**	**Assembler**	**Linker**
	ARMCC	ARMASM	ARMLNK

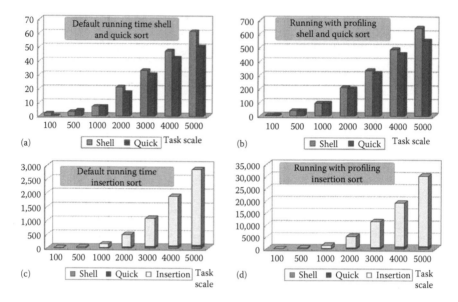

FIGURE 8.6 Experimental results for shell, quick, and insertion sort. X-axis denotes the task scales, while y-axis indicates the running time on different occasions. (a) and (c) present the default running time for the three different sort applications, while (b) and (d) illustrate the running time with the profiling techniques integrated.

scale, the gap between insertion sort and shell/quick sort is also becoming larger, from 9× (task scale is 100) to 42× (task scale is 5,000), respectively.

8.5.1.2 Execution Time with Profiling

Compared to Figure 8.6a and 8.6c, Figure 8.6b and 8.6d presents the task running time under profiling conditions. Due to general instrumentation operations in the software profiling method, the execution time is significantly increased. In our experiment, every single instruction is collected during execution, which leads to a 9.9× ~ 11.0× speedup of the original program execution. Meanwhile, the performance gaps between insertion sort can be achieved by comparing 8.6c and 8.6d, which are similar to the shell/quick sort in 8.6a and 8.6b, from 8.76× to 40.17×.

8.5.2 Hot Spot Location

From the software profiling techniques, the Strcmp function, which is responsible for the string comparing operation, is called with highest frequency by the three sort applications. Figure 8.7 describes the percent of

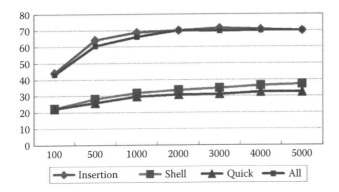

FIGURE 8.7   Percentage of Strcmp function as hot spot.

Strcmp functions called by each sort application as well as the application including the three sorts. The *x*-axis is the task scales, while the *y*-axis is the percentage. For insertion sort, the percentage grows with task scale simultaneously, from 44.14% to 69.99%, while the number in quick sort ranges from 22.09% to 32.37%. Furthermore, the percentage in the whole application is from 43.14% to 70%, which is quite close to the sole insertion sort program. The result is reasonable and acceptable since the insertion sort application takes 96.05% of the execution time approximately when task scale is 5,000.

8.5.3 Profiling Overheads

The profiling overheads are illustrated in Figure 8.8. Adding profiling mechanism will take serious execution overheads, resulting in the execution time with profiling more than 10 times longer than the original

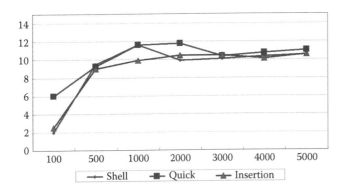

FIGURE 8.8   Profiling overheads.

execution time. The *y*-axis represents speedup that is the execution time under the default execution time against the profiling condition, which is described in Equation 8.1.

$$\text{Speedup} = \frac{T_{\text{Profiling}}}{T_{\text{Original}}} \tag{8.1}$$

When there are fewer tasks, the execution time and additional instrumentation time are proportional to the number of tasks, resulting in a low profiling overhead. In contrast, as the task scale grows bigger, it takes much more time to record the instruction every clock cycle, including the profiling files access. Therefore, speedup achieves approximately 11.0× when task scale is increased to more than 1,000.

8.5.4 Running Mode, Optimization, and Code Size

It is common knowledge that working under Thumb mode significantly reduces the code size. Table 8.3 lists both the code and data sizes working in ARM and Thumb modes.

8.5.4.1 ARM Mode versus Thumb Mode

The code line represents the size of code segment, RO and RW data refer to the read-only and read-write data, and ZI stands for the zero-initialized data. In summary, RO total includes code and RO data, RW total includes RW data and ZI data, and final ROM consists of code, RO data, and RW data. From the table we can get following conclusions:

First, considering the different optimization levels on both ARM and Thumb modes, the code size is O0 > O1 > O2 = O3, while the debug information size is O0 < O1 < O2 < O3. The results prove that O0 has the minimum level for

TABLE 8.3   ARM Mode versus Thumb Mode in Code and Data Size

	Optimization Levels at ARM Mode				Optimization Levels at Thumb Mode			
	O0	O1	O2	O3	O0	O1	O2	O3
Code size	9,840	9,752	9,688	9,688	7,196	7,132	7,108	7,108
RO data	320	320	320	320	320	320	320	320
RW data	28	28	28	28	28	28	28	28
ZI size	556	556	556	556	556	556	556	556
Debug	6,900	6,948	7,386	7,392	7,372	7,456	7,832	7,848
RO total	10,160	10,072	10,008	10,008	7,516	7,452	7,428	7,428
RW total	584	584	584	584	584	584	584	584
ROM	10,188	10,100	10,036	10,036	7,544	7,480	7,456	7,456

optimization, and also contains least debugging information, and vice versa for O3. The debugging information only takes 1.5% of the original code size.

Second, the ARM code size is bigger than Thumb code size. On average, ARM code size is 36.5% bigger than Thumb code, which makes the RO and ROM code sizes 35.0% and 34.8% bigger, respectively. These results are acceptable since Thumb mode only contains 16-bit instructions, while it brings much more instruction numbers.

8.5.5 ARM and Thumb Modes on Different Task Scales

Figure 8.9 illustrates further experimental results of ARM and Thumb modes running with different task scales. The vertical axis indicates the speedup running under ARM mode against Thumb mode, calculated in Equation 8.2.

$$\text{Speedup} = \frac{T_{\text{Thumb}}}{T_{\text{ARM}}} \tag{8.2}$$

The results depict that when the task scale is bigger than 500, the performance of ARM mode is better than that of Thumb mode. In particular, when the task scale is 5,000, the speedup of insertion sort, shell sort, and quick sort are 1.19×, 1.40×, and 1.33×, respectively.

8.5.6 ARM and Thumb Modes on Optimization Levels

Figure 8.9b presents the experimental results of ARM and Thumb modes running at different optimization levels. In this test case, we set the task scale to 5,000 to measure the leverage among different optimization levels.

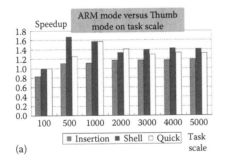 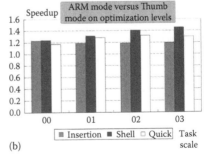

FIGURE 8.9   ARM mode versus Thumb mode in different task scales and optimization levels. (a) Illustrates the speedup on different task scales, while (b) presents the speedup with different optimizing options.

The experimental speedup of insertion sort reaches 1.23X at O0, while the shell sort and quick sort achieve 1.45× and 1.3× at O3, respectively.

8.6 CONCLUSIONS

In this chapter we have proposed Codem, a novel high-level modeling approach and design flow for reconfigurable coprocessor architectures on FPGA platform. Abstracted from a three-layer hierarchical model, Codem provides an adaptive mapping method with unified programming interfaces and services integration interfaces under reconfigurable condition. Furthermore, Codem incorporates a software/hardware codesign flow with profiling techniques. Empirical results using sort applications demonstrate that Codem can integrate multiple types of IP accelerators under reconfigurable condition in a flexible manner. From the experimental results we can also get the conclusion that by integrating SOA concept to MPSoC architecture design, flexibility and efficiency of MPSoC have been improved so that the design complexity of the prototype system is reduced, and TTM also is shortened at the same time.

There are numerous future directions worth pursuing. First, improved task partition and further adaptive mapping schemes will be important to efficiently support automatic task level parallelization. Second, we also plan to study about the out-of-order task execution paradigm, exploring potential parallelism of sequential programs. Finally, we believe that Codem opens a new research direction of utilizing the hardware service concepts to efficiently manage heterogeneous chips at a higher level of abstraction.

FUNDING

This work was supported by the National Science Foundation of China [61379040], [61272131], and [61202053]; Jiangsu Provincial Natural Science Foundation [SBK201240198]; Open Project of State Key Laboratory of Computer Architecture Institute of Computing Technology Chinese Academy of Sciences [CARCH201407]; and the Strategic Priority Research Program of CAS [XDA06010403].

REFERENCES

S. Borkar, A. A. Chien. The future of microprocessors[J]. *Commun. ACM*:2011, 54(5): 67–77.

R. F. Cmelik. *SpixTools: Introduction and User's Manual*[R]. Sun Microsystems, Menlo Park, CA, 1993.

K. Compton, S. Hauck. Reconfigurable computing: A survey of systems and software[J]. *ACM Comput. Surv.*:2002, 34(2): 171–210.

D. Gohringer, M. Hubner, V. Schatz et al. Runtime adaptive multi-processor system-on-chip: RAMPSoC. *IEEE International Symposium on Parallel and Distributed Processing*, Miami, FL, 2008.

S. C. Goldstein, H. Schmit, M. Budiu et al. PipeRench: A reconfigurable architecture and compiler[J]. *Computer*:2000, 33(4): 70–77.

S. Hauck, T. W. Fry, M. M. Hosler et al. The chimaera reconfigurable functional unit[J]. *IEEE Trans. Very Large Scale Integr. Syst.*:2004, 12(2): 46–57.

J. R. Hauser, J. Wawrzynek. Garp: A MIPS processor with a reconfigurable coprocessor. *Proceedings of the 5th Annual IEEE Symposium on FPGAs for Custom Computing Machines*, Napa, CA, 1997.

D. Koch, C. Beckhoff, J. Teich. A communication architecture for complex runtime reconfigurable systems and its implementation on spartan-3 FPGAs. *Proceedings of the ACM/SIGDA International Symposium on Field Programmable Gate Arrays*, Monterey, CA, ACM, 2009.

G. Kuzmanov, G. Gaydadjiev, S. Vassiliadis. The MOLEN processor prototype. *Proceedings of the 12th Annual IEEE Symposium on Field-Programmable Custom Computing Machines*, Napa, CA, 2004.

E. Lubbers, M. Platzner. ReconOS: Multithreaded programming for reconfigurable computers[J]. *ACM Trans. Embedded Comput. Syst.*:2009, 9(1): 1–33.

L. M. Mourelle, R. E. Ferreira, N. Nedjah. Migration selection of strategies for parallel genetic algorithms implementation on NoCs[J]. *Int. J. Electron.*:2010, 97(10): 1227–1240.

N. Nedjah, L. De Macedo Mourelle. A hardware/software co-design versus hardware-only implementation of modular exponentiation using the sliding-window method[J]. *J. Circ Syst Comput*:2009, 18(2): 295–310.

C. Neely, G. Brebner, W. Shang. ShapeUp: A high-level design approach to simplify module interconnection on FPGAs. *Proceedings of the 18th IEEE Annual International Symposium on Field-Programmable Custom Computing Machines*, IEEE Computer Society, Charlotte, NC, 2010.

W. Peck, E. Anderson, J. Agron et al. *Hthreads: A Computational Model for Reconfigurable Devices International Conference on Field Programmable Logic and Applications*, Madrid, Spain, 2006.

K. Rupnow, K. D. Underwood, K. Compton. Scientific application demands on a reconfigurable functional unit interface[J]. *ACM Trans. Reconfigurable Technol. Syst.*:2011, 4(2): 1–30.

M. V. C. Silva, N. Nedjah, L. M. Mourelle. Power-aware multi-objective evolutionary optimization for application mapping on NoC platforms[J]. *Int. J. Electron.*:2010, 97(10): 1163–1179.

S. Singh. Computing without processors[J]. *Communications of ACM*:2011, 54(8): 46–54.

D. Tarditi, S. Puri, J. Oglesby. Accelerator: using data parallelism to program GPUs for general-purpose uses. *Proceedings of the 12th International Conference on Architectural Support for Programming Languages and Operating Systems*, San Jose, CA, ACM, 2006.

C. Wang, X. Li, J. Zhang et al. FPM: A flexible programming model for MPSoC on FPGA. *Proceedings of the 19th Reconfigurable Architecture Workshop*, Shanghai, China, IEEE Computer Society, 2012.

C. Wang, X. Li, J. Zhang et al. MP-Tomasulo: A dependency-aware automatic parallel execution engine for sequential programs[J]. *ACM Trans. Archit. Code Optim.*:2013, 10(2): 1–24.

C. Wang, J. Zhang, X. Zhou et al. SOMP: Service-oriented multi processors. *Proceedings of the IEEE International Conference on Services Computing*, IEEE Computer Society, Washington, DC, 2011.

M. A. Watkins, D. H. Albonesi. ReMAP: A reconfigurable heterogeneous multi-core architecture. *Proceedings of the 43rd Annual IEEE/ACM International Symposium on Microarchitecture*, IEEE Computer Society, 497–508, Atlanta, GA, 2010.

J. Wawrzynek, D. Patterson, M. Oskin et al. RAMP: Research accelerator for multiple processors[J]. *Micro. IEEE*:2007, 27(2): 46–57.

D. Wentzlaff, A. Agarwal. Factored operating systems (fos): The case for a scalable operating system for multicores[J]. *ACM SIGOPS Operating Systems Review*:2009, 43(2): 76–85.

R. D. Wittig, P. Chow. OneChip: An FPGA processor with reconfigurable logic. *Proceedings of the IEEE Symposium on FPGAs for Custom Computing Machines*, Napa, CA, 1995.

CHAPTER 9

Greedy Partitioning and Insert Scheduling Algorithm for Hardware–Software Codesign on MPSoCs

Chao Wang, Chunsheng Li, Xi Li, Aili Wang, Fahui Jia, Xuehai Zhou, and Nadia Nedjah

CONTENTS

9.1 INTRODUCTION

The tremendous invasion of multiprocessor system-on-chip (MPSoC) has brought numerous computation abilities to heterogeneous platforms in the past decades. However, it still poses significant challenges to partition and task scheduling to different function units, especially for tasks with data dependencies. Task partitioning and scheduling problems are intractable in many applications by its non-deterministic polynomial-complete characters [1–2], especially in hardware–software (HW–SW) codesign. It is well known that efficient partitioning and scheduling algorithms have a central impact on global performance improvement, for instance, energy, power, area, and acceleration. MPSoC, which has emerged for decades, is now dominating and will eventually become the pervasive computing model. It has both hardware and software cores and needs high-performance task partitioning and scheduling algorithms.

The object of partitioning is to decide whether the task should be implemented in hardware or software. Generally, software implementation is flexible and sequential in execution; in contrast, hardware implementation is fixed and parallel. Hence, performance- or power-critical tasks of the system should be realized in hardware, while noncritical components can be done in software; alternatively, in this way, an optimal trade-off among cost, power, and performance can be achieved [3]. The aim of scheduling is to minimize the overall execution time of the parallel applications by properly allocating and rearranging the execution order of the tasks on the cores without violating the precedence constraints among the tasks [4–5].

In order to attack the above-mentioned problems, this chapter proposes a task partitioning and scheduling method that can access an efficient utilization with polynomial time. We claim the following contributions:

1. Combine partitioning and scheduling together with both of their advantages. Take into account the critical paths and the scattered tasks with greedy strategy, then just simply insert operations in the basic orderly queue.

2. An $O(V + E)$ polynomial time complexity with facile computations; most of them can be operated at initialization, and need a little space complexity.

3. Computation cost, area constraint, and communication ratio are all considered simultaneously, while the algorithm has a good scalability for large-scale problems and other MPSoC platforms.

The rest of this chapter is organized as follows: Related work can be seen in Section 9.2. Section 9.3 gives some statements, utilizing system model. In Section 9.4, we discuss partitioning and scheduling algorithms. Illustrative examples will be presented in Section 9.5. Section 9.6 shows the simulation experiment results and analysis. Finally Section 9.7 draws the work conclusion and future work.

9.2 RELATED WORK

There have been many state-of-the-art research works reported in this field, focusing on different aspects in the HW–SW partitioning and task scheduling. Traditional HW–SW partitioning approaches include software oriented [6] and hardware oriented [7]. The distinction between the two methods depends on which HW–SW is initialized first and iteratively moving to the other HW–SW with the performance constraints. Many approaches pay attention to the algorithm aspects, that is, accurate algorithms, including Dynamic Programming [8], Integer Linear Programming [9], Branch and Bound [10], which suit small-scale problems; other heuristic algorithms like Genetic Algorithms [11], Simulated Annealing [12], Tabu Search [13], and greedy strategy [14] are more proper for large-scale questions. Most of these algorithms are based on static strategy; there are also some dynamic methods [15–16].

Partitioning has a close relationship with scheduling and they are problem- or architecture-dependent [17]. In particular, many researchers consider scheduling as a part of partitioning [14,18], whereas others do not [19]. Recently, a trend in HW–SW codesign on MPSoC combines partitioning with scheduling together. Scheduling First Partitioning Later (SFPL) is one method [4,5,20]; Youness and Hassan et al. [4] uses the A-star algorithm for scheduling dependent tasks onto homogeneous processors, then chooses the longest schedule length for hardware implementation, and, the time complexity is $O(p(V3-V2)/2)$. Based on this approach, the channel conflict is handled by graph coloring technique in [5]. The time complexity of this algorithm is large and no area constraint is considered [4–5]. Here, the A-star algorithm is an extension of Dijkstra algorithm, which achieves better performance by using heuristics with cost functions. Hong-lei et al. [20] improve the A-star algorithm time complexity to $O(pV2)$ in scheduling and introduce benefit-to-area ratio as the priority in partitioning; cost functions in the A-star algorithm still need to be calculated many times. Another method is Partitioning First Scheduling Later (PFSL) [17,21], in which three heuristic search partitioning methods are compared with each other [21]; the result shows Tabu Search is the best of the three. Wang et al. [22] have introduced a task out-of-order scheduling engine at task level for MPSoC; it is able to detect the intertask data dependence for automatic parallelization. MP-Tomasulo [23] can not only detect the data dependence, but also has the ability to rename the parameters in order to eliminate the name dependence. Jigang et al. [17] introduce the new benefit function for partitioning, and then use critical-path and communication combined scheduler (CPCS) algorithm for scheduling. Both of the papers choose hardware-implementation tasks like breadth first search (BFS). The hardware nodes are always scattered, not in a full deeply path; meanwhile both the hardware/software implementation is just in one core. It is not a good way for dependent tasks especially when there are large frequent communication times. Li et al. [24] present a scheduling algorithm considering the reconfigurable hardware logic resources on field programmable gate arrays (FPGAs).

The advantages of SFPL and PFSL are combined in this chapter: deeply critical or longest paths are calculated by SFPL, while scattered hardware/software nodes are found by PFSL and inserted with greedy scheduling. We name this approach greedy partitioning and insert scheduling method, detailed in Sections 9.4 and 9.5.

9.3 TARGET SYSTEM AND GRAPH MODEL

9.3.1 MPSoC Architecture

Today, there are variety of MPSoC architecture and platforms for different purposes. The target system architecture in this chapter is illustrated in Figure 9.1, based on Xilinx FPGAs. There is a main control CPU in charge of task partitioning and scheduling; it deals with the generated task graph into software or hardware cores. Software cores and hardware logic unit communicate via bus connection, while each core (software or hardware) has its own local memory (LM) for intertask communication destined in the same core. In order to provide data communications between processors, one shared memory block is implemented.

Xilinx FPGAs are the most widely used programmable silicon foundation for targeted design platforms, which deliver integrated software and hardware components. For the ease of the IP-based modular and scalable architecture design, Xilinx FPGA supplies PowerPC, MicroBlaze/PicoBlaze

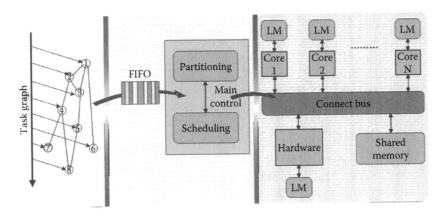

FIGURE 9.1   MPSoC architecture constructed in Xilinx FPGA.

cores, and advanced DSP slices in one chip, and distribute RAM and block RAM are used for memory storage. The utilization of hardware resources, like configurable logic blocks (CLBs) for FPGA in particular, plays a vital role during the evaluation and metrics of programmable devices. In this chapter, we select Xilinx FPGA as the foundation of abstract architecture platform and measure the equipped CLBs for area constraint evaluations.

Throughout this chapter, we define the term *task* as a coarse-grained set of computation or function unit, and pick up a static strategy for partitioning and scheduling. Tasks are modeled as nodes in the graph and can be paralyzed either in hardware implementation or sequentially in software implementation. We put the limitations into assumption preliminaries as follows:

1. Each selected task in the task graph can be virtualized only to one specific core; the selection of homogeneous/heterogeneous core is according to the task performance.

2. Tasks can be implemented either by software or by hardware. Software can handle a single task at a time by sequential execution, while hardware handles multitasks by paralyzed execution according to the area constraint of the system.

3. Once the task starts execution, it cannot be interrupted.

4. Connect bus has enough bandwidth for handling the conflicts.

5. Communication time is the total time including read, write, and so on.

6. Communication between tasks in one core is cost free.

9.3.2 Task Graph Model

The task graph model is generated by the data flow graph (DFG), and usually expressed as the directed acyclic graph (DAG). In this chapter we give the six-group expression as follows:

$$G=(V|(Sw,Hw,A),\ E|C)$$

where:

  V: The set of nodes in DAG, which represent dependent tasks

  Sw: The execution time of tasks (node V) by software

  Hw: The execution time of tasks (node V) by hardware

  A: The area consumption of tasks implemented by hardware

  E: The set of edges in DAG, which refers to intertask data communication between tasks

  C: The communication time between two tasks

Based on the definition task model, a sample of task graph is presented in Figure 9.2.

The demonstrative task graph example in Figure 9.2 is originated from input task V_I and ends up with the final output task V_O. Between the input and output tasks, 11 tasks are generated, and each of them is illustrated in the square box. The top part of the task refers to the ID assigned during initialization, while the bottom part indicates the tuples including Sw, Hw, and A. The digit on each edge represents the communication overheads between the two tasks connected by the edge.

In particular, we define the following terms:

Definition 9.1:

V refers to the tasks set $\{V_1, V_2, ..., V_n\}$. Considering all the nodes in DAG, if there exists a path between two nodes, which is presented as $V_i \xrightarrow{C(V_i,V_j)} V_j$ in DAG, V_i is the predecessor of V_j; meanwhile V_j is the successor of V_i.

$$\begin{cases} P\_Set(V)=\{V'|(V',V)\in E\} \\ S\_Set(V)=\{V'|(V,V')\in E\} \end{cases} \qquad (9.1)$$

The value of $C_r (V_i, V_j)$ denotes the *real* communication time between two dependent tasks.

$$C_r(V_i,V_j)=\begin{cases} 0, & \text{if } V_i,V_j \text{ on same core} \\ C(V_i,V_j), & \text{if } V_i,V_j \text{ on different cores} \end{cases} \qquad (9.2)$$

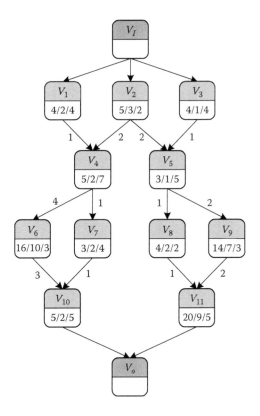

FIGURE 9.2   Task graph example.

Definition 9.2:

A task V_i can be executed if all the tasks in $P\_Set(V_i)$ are finished and the data communication is ready, so the finish time of task V_i is:

$$T_f(V_i) = \begin{cases} \max\{T_f[P\_Set(V_i)] + C_r[P\_Set(V_i), V_i]\} + Sw(V_i), \\ \quad \text{if } V_i \text{ is a software task} \\ \max\{T_f[P\_Set(V_i)] + C_r[P\_Set(V_i), V_i]\} + Hw(V_i), \\ \quad \text{if } V_i \text{ is a hardware task} \end{cases} \qquad (9.3)$$

Definition 9.3:

The virtual input task V_I and output task V_O are added into the set of tasks for the convenience of DAG algorithms.

$$V_I : \begin{cases} P\_Set(V_I)=\varnothing \\ C[V_I,S\_Set(V_I)]=0 \\ T_f(V_I)=0 \end{cases} \tag{9.4}$$

$$V_O : \begin{cases} S\_Set(V_O)=\varnothing \\ C[P\_Set(V_O),V_O]=0 \end{cases}$$

According to the definitions, $T_f(V_O)$ indicates the time when all the tasks (nodes in DAG) are finished. Hence the design object of task partitioning and scheduling is to find a minimum of $T_f(V_O)$.

9.4 ALGORITHM

9.4.1 Greedy Partitioning

9.4.1.1 Benefit Function

For each task (node in DAG), following characters are taken into consideration: hardware execution time, software execution time, and area constraints. All these factors are limited by the utility of graph theory. In order to leverage the influences of the parameters, we combine these factors into benefit-to-area function [20].

Definition 9.4:

For each task V in DAG, the benefit-to-area function is defined as follows:

$$B(V)=\frac{Sw(V)-Hw(V)}{A(V)} \tag{9.5}$$

$B(V)$ denotes the time saved on the unit area. We use subtraction instead of division (acceleration) because the time subtraction on the unit area not only reflects saved time, but also reflects the area constraint. The benefit value can be calculated at beginning and the time complexity is $O(V)$.

9.4.1.2 Critical Hardware Path

To identify the critical hardware path, our object is to find an orderly path from V_I to V_O, in which the sum of tasks' benefits is max. On this basis, the area occupation constraints of these tasks fitting global hardware constraint is defined as follows:

Definition 9.5:

The critical hardware path is an orderly path in the DAG, in which the sum of tasks' benefit-to-area values achieves the peak, while the total area occupation of tasks in this candidate path does not exceed the global area limitations. An orderly list is presented in formula 9.6:

$$Hw_{order} : \left\{ V_I \to V_i \to V_j \to \cdots \to V_O \middle| \max \sum B(V_i) \text{ and } \sum A(V_i) \le A_{all} \right\} (9.6)$$

DAG is updated by removing nodes and edges between them except V_I and V_O, and then new critical path will be searched if it exists in the updated DAG, which means this operation can be done repeatedly by searching and updating. Otherwise, some scatter nodes can be found just according to area occupation and ordered by benefit value, which is called *hard-like* nodes:

$$Hw_{like} : \left\{ \begin{array}{l} V_j \middle| A(V_j) \le A_{all} - \sum A_{path} \\ V_i, V_j \text{ if } A(V_i) < A(V_j) \text{ or } B(V_i) > B(V_j) \middle| A(V_i) = A(V_j) \end{array} \right\} (9.7)$$

These *hard-like* nodes are not removed in the updated DAG.

9.4.1.3 Longest Software Path

We use communication time (edge value) to find the longest software path in the updated DAG from Section 9.4.1.2. Similar to the hardware paths, a software path is an orderly line too.

Definition 9.6:

The longest software path is an orderly path in the DAG; the sum of communication time between nodes in this path is the biggest among other paths:

$$Sw_{order} : \left\{ V_I \to V_m \to V_n \to \cdots \to V_O \middle| \max \sum C(V_i) \right\} (9.8)$$

If this path has some *hard-like* nodes like $V_m \to V_{Hwlike} \to V_n$, a comparison is operated to choose the path or node options.

$$\left\{ \begin{array}{l} C(V_m, V_{Hwlike}) + Hw(V_{Hwlike}) + C(V_{Hwlike}, V_n) > Sw(V_{Hwlike}), \\ \text{keep path and remove node} \\ C(V_m, V_{Hwlike}) + Hw(V_{Hwlike}) + C(V_{Hwlike}, V_n) \le Sw(V_{Hwlike}), \\ \text{remove path and keep node} \end{array} \right. (9.9)$$

We may get multiple longest software paths by updating DAG, the *hard-like* nodes in Hw_{like} are renewed by formula 9.9 operations. Furthermore, when there is no complete path from V_I to V_O in the updated DAG, add the rest nodes in the set:

$$Sw_{rest} : \left\{ V_n \middle| V_n \in V - Hw_{order} - Hw_{like} - Sw_{order} \right\} \tag{9.10}$$

The key step of this algorithm is to find the longest path in DAG. The method is similar to the critical path exploration in activity on edge network; therefore the time complexity of the path exploration is $O(V+E)$. Due to the largest comparing time of $O(V)$, the total time complexity of orderly partitioning is $O(V+E) + O(V+E) + O(V) = O(V+E)$.

9.4.2 Insert Scheduling

Four sets of tasks (Section 9.4.1) are further divided into two categories: paths set (may be more than one path) and nodes set. According to Definition 9.2, the task's order in these paths cannot be modified afterward. Single node can be inserted into these orderly paths at a proper position to shorten the total execution time as much. To be specific, following criteria are taken into account:

1. Hardware tasks can be executed in parallel.

2. Every node can enter execution stage once all of its parent nodes are finished.

3. Software tasks can be put in multiple cores indicated by longest edge path.

4. Scatter software tasks can be put in an independent core if there are enough available cores and no tasks are being executed simultaneously.

5. The node that has more successor hardware tasks will be appointed to a higher priority of execution, comparing with the same level nodes before insertion.

6. When there are similar tasks at the same level, the one with the shortest execution time should be executed first.

The purpose of insert scheduling is to find an optimal execution time and maintain the overall system utilization; the scatter tasks can be inserted into an orderly path (already existed core) or just be kept into a new independent core. The numbers of needed cores are added according to platform constraints and insert criteria mentioned earlier. The upper bound of cores is in line with conclusion of Youness et al. [4] that scheduling length of optimal task assignment to $P + 1$ processors is always no bigger than the one to P processors. The time complexity of insert scheduling is $O(V)$, because only some scatter nodes need to be compared, and the maximum of compared nodes is V.

9.4.3 Algorithm Flowchart

The algorithm flowchart is described in Figure 9.3. First, the benefit value of each task is calculated as the reference for task partitioning, and the critical path, which has maximum sum of tasks' benefit values and total area occupation fitting the requirements, will be chosen. Meanwhile the global area constraint is updated by subtracting this critical path area occupation, and the DAG is also updated by removing nodes and edges between them in this critical path. Then the greedy strategy is used for rest paths until all the *hard-like* nodes are found. Second, the longest path and rest nodes will be found by the same way; meanwhile *hard-like* nodes will be renewed according to comparison with computation and communication time. Finally the orderly tasks in the critical or longest path will be assigned to alternative hardware or software core, and scatter nodes are inserted according to insert scheduling criterions.

9.5 ILLUSTRATIVE EXAMPLE

In order to demonstrate the effectiveness of our proposed algorithm, we run a test case in Figure 9.4, similar to the one presented by Jigang et al. [17]. For better describing the algorithm, the nodes V_I and V_O are omitted for simplicity.

Step 1: Calculate every task's benefit-to-area value as Figure 9.4a.

Step 2: Find the critical hardware path $V_2 \rightarrow V_5 \rightarrow V_9 \rightarrow V_{11}$; the sum of area occupation is 15 (global area constraint is 18); there are no other critical paths in the updated DAG, and so we get:

$$Hw_{order} \left\{ V_2 \rightarrow V_5 \rightarrow V_9 \rightarrow V_{11} \right\}$$

Step 3: Find the *hard-like* nodes set according to formula 9.7; the area occupation of V_8 is smaller than V_6, so the order is:

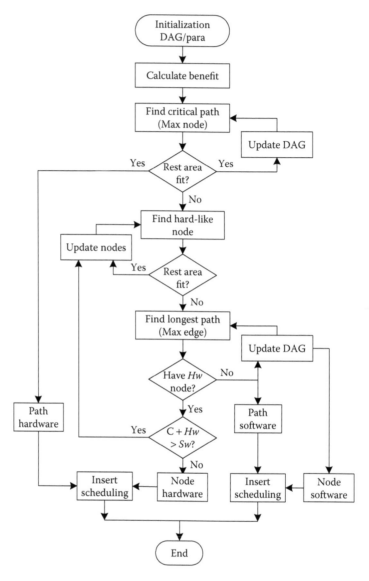

FIGURE 9.3 Algorithm flowchart.

$$Hw_{like}\{V_8, V_6\}$$

Step 4: Updated DAG as Figure 9.4b, and then find the longest software path $V_1 \rightarrow V_4 \rightarrow V_6 \rightarrow V_{10}$; V_8 is not in this longest software path, so V_8 can be made by hardware. As there is not enough area for V_6, renew *hard-like* nodes set as follows:

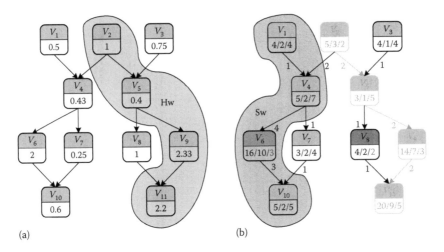

FIGURE 9.4   Illustrative example—DAG graph. (a) Calculate benefit-to-area value and (b) Updated DAG.

$$Hw_{like}\{V_8\}$$

If the area occupation of V_8 is assumed to be 3, we can get *hard-like* nodes set $Hw_{like}\{V_6,V_8\}$ by formula 9.7. There is a node V_6 in the set of Hw_{like}, according to formula 9.9, $4 + 10 + 3 > 16$, so keep this longest path and V_6 should be removed from *hard-like* nodes set.

There is no other longest path in the updated DAG; by removing found longest software path $V_1{\rightarrow}V_4{\rightarrow}V_6{\rightarrow}V_{10}$, we can get:

$$Sw_{order}\{V_1 \rightarrow V_4 \rightarrow V_6 \rightarrow V_{10}\}$$

Step 5: Find the rest nodes set:

$$Sw_{rest}\{V_3,V_7\}$$

The above steps are greedy partitioning as shown in Figure 9.4. Shaded area in Figure 9.4a is the critical hardware path, while the longest software path is shown in Figure 9.4b. The node V_6 (dark in b) is the *hard-like* node in the temp calculation process and finally is deleted from the set of *hard-like* nodes. The node V_8 (dark in b) is the end result of *hard-like* node.

The rest steps are insert scheduling, and the results can be represented by Gantt chart.

Step 6: Insert V_8 into the hardware implementation set, which can be executed in parallel with V_9 as soon as V_5 is completed. Tasks in the longest path (Sw_{order}) are assigned into one software core, while the rest of software nodes are assigned into another core, like Gantt chart in Figure 9.5.

Step 7: If there is only one software core, all the tasks that are implemented by software should be inserted into this core. According to criterion 5, the successor of V_3 is hardware, so V_3 should be executed first before V_1, whose successor is software; V_7 should be executed before V_6 according to criterion 6. The procedure is presented in Figure 9.6.

The whole execution time is 31 ns in Figure 9.5 (two software cores) and 37 in Figure 9.6 (one software core). Taking the CPCS strategy [17] into consideration, the hardware-implementation tasks are V_2, V_6, V_8, V_9, and V_{11}; in this case the whole execution time is 42, and it needs more calculations as well as longer communication time.

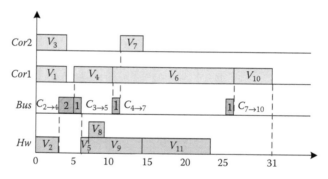

FIGURE 9.5 Illustrative example—Gantt chart (2 cores).

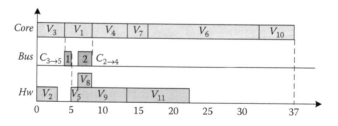

FIGURE 9.6 Illustrative example—Gantt chart (1 core).

9.6 EXPERIMENTAL RESULTS AND ANALYSIS

There is a broad range of applications with different scheduling and partitioning algorithms. The assumptions, preconditions, and benchmarks vary from each other. So it is hard to tell which application-oriented algorithm has a better performance. The evaluation of the algorithm performance in this chapter depends on the platform configuration and application utilization presented [17,21].

DAGs used in this chapter are randomly generated task graphs with a uniform distribution and commonly encountered structure: in-tree, out-tree, fork-joint, mean value analysis, and FFT (Figure 9.7). These five kinds of DAGs are randomly generated by task graphs for free (TGFF) [25], which provides a flexible and standard way of generating pseudo-random task graphs during the scheduling and allocation research.

There are some parameters of DAGs listed in Figure 9.2, which are defined as follows:

The last two rows in the table are the different communication time cases: in case (1), the upper part is similar to the presentation in Stitt et al. [16] and Hong-lei et al. [20], while in case (2) the bottom part improves the communication cost in the total execution time. In contrast to the *hardware task in percentage* definitions in Stitt et al. [16], we define *acceleration ratio (AR)* and *area percentage (AP)* as follows:

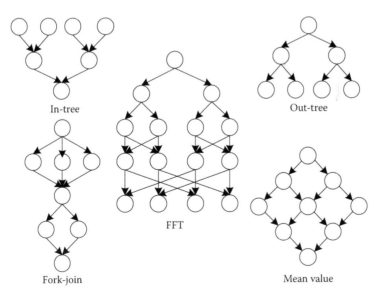

FIGURE 9.7   Five kinds of task graphs (DAGs).

Definition 9.7:

Assume there are n tasks in the DAG, denoted as V_i ($i = 1:n$), and every task occupies $A(V_i)$ unit area. The global area constraint takes p percentage of all tasks' area occupation, which is defined as AP (area percentage):

$$AP = \left[p \times \sum_{i=1}^{n} A(V_i) \right] \left(p = \frac{1}{5}, \frac{2}{5}, \frac{1}{3}, \frac{1}{2}, \frac{3}{5}, \frac{2}{3}, \frac{4}{5} \right) \tag{9.11}$$

Definition 9.8:

If only one software core is available in the platform, we assume that all the tasks are implemented by software, and then the communication time is free. In this situation the longest execution time is defined as the sum of all software time, and the AR (acceleration ratio) for one software core is:

$$AR_{one} = \frac{\text{Assume longest time}}{\text{Algorithm time}} = \frac{\sum_{i=1}^{n} Sw(V_i)}{\text{Algorithm time}} \tag{9.12}$$

Definition 9.9:

If there is more than one software core used in the platform, we assume DAG has l levels and the communication time is not free. So the longest execution time is defined as the sum of maximum of software time and communication time level by level. The AR for multi-software cores is:

$$AR_{multi} = \frac{\sum_{i=1}^{l} \text{Max}\{Sw(V_i), C(V_i) \text{ in one level}\}}{\text{Algorithm time}} \tag{9.13}$$

Take Figure 9.2 as an illustrative example, where the AP is $18 = [2/5 \times$ all area]. The AR is $83/37 = 2.2432$ for single software core situation, while $55/31 = 1.7741$ for a multicore scenario.

In this chapter, the output is described by AR not by scheduling length because results of scheduling length are dominantly affected by random parameters under different conditions, which cannot reflect the performance of the algorithm.

The input is random DAGs with dependent tasks by TGFF; the parameters such as software time, hardware time, area occupation, and communication time are randomly generated according to Table 9.1. We use average of 30×5 (5 kinds of DAGs in Figure 9.7) experiments by computer simulation.

Figure 9.8 depicts the comparison between the GPISM with the state-of-the art CPCS. The chart is based on the communication time case (1). X-axis in the figure represents AP, while Y-axis refers to the change of AR according to AP. The numerical result of Figure 9.8a is slightly higher than that of Figure 9.8b, because Figure 9.8a is single software core and so all the software tasks should be executed in sequence; when the tasks are changed into hardware implementation, the results should be higher obviously. These results have an upper bound to reach the theoretical speedup, which is the division between the time that all tasks are implemented by software execution time and hardware execution time of all the tasks. From Figure 9.8, we can see that the performance of these two algorithms is very close to each other, because the proportion of communication time is less than calculation time, but GPISM needs less computation workloads than CPCS method with similar performance.

TABLE 9.1   Parameters Defined in DAGs

Parameters of DAG	Values
Range of software execution time	800 ~ 2000 (ns)
Range of hardware execution time	200 ~ 1200 (ns)
Range of hardware area	100 ~ 400 (unit)
Range of communication time[1]	2 ~ 100 (ns)
Range of communication time[2]	20 ~ 800(ns)

Note: (1) and (2) are the case studies

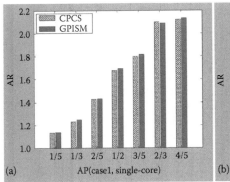

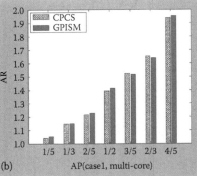

FIGURE 9.8   AR with AP (case 1). (a) Acceleration ratio on Software Execution with Single Core and (b) Acceleration ratio on Hardware Execution with Multicore.

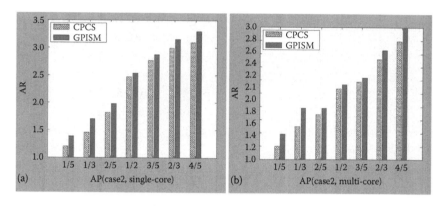

FIGURE 9.9   AR with AP (case 2). (a) Acceleration ratio on Software Execution with Single Core and (b) Acceleration ratio on Hardware Execution with Multicore.

Furthermore, Figure 9.9 presents the AR changes with communication time case (2), and simulates runtime comparison with two algorithms in the same configuration and environment. The overall performance increase ratio of GPISM than CPCS is about 15% in our average of 30×5 experiments with specific parameters. As the DAGs and their parameters are randomly generated, maybe this result changes according to different conditions, such as random communication cost by statistical information. We believe that when the communication cost scales, the AR with GPISM is larger than CPCS algorithm, because the latter searched for scattered tasks, which may lead to more communication time than the former one. Meanwhile, the performance of GPISM still maintained at a stable level with the expansion of the scale of the problem for its efficient utilization. GPISM makes leverage between algorithm complicity and system performance, and this algorithm can be extended to complex MPSoC environments.

9.7 CONCLUSIONS AND FUTURE WORK

In this chapter, we have proposed a highly efficient greedy partitioning and insert scheduling method on hardware–software partition problem. The timing complexity of the proposed approach is a polynomial time $O(V + E)$. Meanwhile the algorithm can be applied to the situation with considerable communication cost and the performance keeps stable. Simulation results on sample applications demonstrate that the algorithm makes a trade-off between computation complexity and optimal results.

Although the platform architecture in this chapter has only one hardware zone according to Xilinx FPGA limitations, we believe that the method can be extended to multiple hardware cores with changeable communication configurations.

In spite of the promising results, there are a lot of directions worth pursuing. On one hand, random DAGs with different parameters have a great impact on the performance of algorithms, and rigorous experimental design and scientific evaluation are needed for further exploration and research; on the other hand, since the reconfigurable heterogeneous MPSoC has been regarded as one of the major trends in the future, our next step is to extend this method to reconfigurable FPGA, in order to take benefits of runtime partial reconfiguration features and technical supports from FPGA research communities.

FUNDING

This work was supported by the National Science Foundation of China [61379040], [61272131], and [61202053]; Jiangsu Provincial Natural Science Foundation [SBK201240198]; Open Project of State Key Laboratory of Computer Architecture; Institute of Computing Technology, Chinese Academy of Sciences [CARCH201407]; and the Strategic Priority Research Program of CAS [XDA06010403].

REFERENCES

1. M. R. Garey and D.S. Johnson. *Computers and intractability: A guide to the theory of NP-completeness*, W.H. Freeman, San francisco, CA, 1979.
2. R. L. Graham. Bounds on multiprocessing timing anomalies. *SIAM Journal on Applied Mathematics*, 1969. **17**: pp. 416–429.
3. J. Wu, T. Srikanthan et al. Algorithmic aspects of hardware/software partitioning: 1D search algorithms. *IEEE Transactions on Computers*, 2010. **59**(4): pp. 532–544.
4. H. Youness, M. Hassan et al. A high performance algorithm for scheduling and hardware-software partitioning on MPSoCs. In *Proceedings of the 4th International Conference on Design & Technology of Integrated Systems in Nanoscal Era*, Cairo, Egypt, 2009: pp. 71–76.
5. H. Youness, A. M. Wahdan et al. Efficient partitioning technique on multiple cores based on optimal scheduling and mapping algorithm. In *Proceedings of IEEE International Symposium on Circuits and Systems* Paris, France, 2010: pp. 3729–3732.
6. F. Vahid and D. D. Gajski. Clustering for improved system-level functional partitioning. In *Proceedings of the 8th International Symposium on System Synthesis*, Cannes, France, 1995: pp. 28–33.

7. R. Niemann and P. Marwedel. Hardware/software partitioning using integer programming. In *Proceedings of the European Design and Test Conference*, Paris, France, 1996: pp. 473–479.

8. J. Wu and T. Srikanthan. Low-complex dynamic programming algorithm for hardware/software partitioning. *Information Processing Letters*, 2006. **98**(2): pp. 41–46.

9. K. Shiann-Rong, C. Chin-Yang et al. Partitioning and pipelined scheduling of embedded system using integer linear programming. In *Proceedings of the 11th International Conference on Parallel and Distributed Systems*, Denver, CO, 2005: pp. 37–41.

10. K. S. Chatha and R. Vemuri. Hardware-software partitioning and pipelined scheduling of transformative applications. *IEEE Transactions on Very Large Scale Integration (VLSI) Systems*, 2002. **10**(3): pp. 193–208.

11. Z. Yi, Z. Zhenquan et al. HW-SW partitioning based on genetic algorithm. In *Congress on Evolutionary Computation*, Portland, OR, Vol. 1, 2004: pp. 628–633.

12. L. Lanying, S. Yanbo et al. A new genetic simulated annealing algorithm for hardware-software partitioning. In *Proceedings of the 2nd International Conference on Information Science and Engineering*, Hangzhou, China, 2010: pp. 1–4.

13. L. Lanying and S. Min. Software-hardware partitioning strategy using hybrid genetic and tabu search. In *Proceedings of the International Conference on Computer Science and Software Engineering*, Wuhan, Hubei, 2008: pp. 83–86.

14. K. S. Chatha and R. Vemurl. MAGELLAN: Multiway hardware-software partitioning and scheduling for latency minimization of hierarchical control-dataflow task graphs. In *Proceedings of the 9th International Symposium on Hardware/Software Codesign*, Copenhagen, Denmark, 2001: pp. 42–47.

15. F Le-jun, L. Bin et al. An approach for dynamic hardware /software partitioning based on DPBIL. In *Proceedings of the 3rd International Conference on Natural Computation*, Haikou, China, 2007: pp. 581–585.

16. G. Stitt, R. Lysecky et al. Dynamic hardware/software partitioning: A first approach. In *Proceedings of the Design Automation Conference*, Anaheim, CA, 2003: pp. 250–255.

17. W. Jigang, T. Srikanthan et al. Algorithmic aspects for functional partitioning and scheduling in hardware/software co-design. *Design Automation for Embedded Systems*, 2008. **12**(4): pp. 345–375.

18. M. Lopez-Vallejo and J. Carlos Lopez. On the hardware-software partitioning problem: System modeling and partitioning techniques. *ACM Transactions on Design Automation of Electronic Systems*, 2003. **8**(3): pp. 269–297.

19. F. Vahid. Partitioning sequential programs for CAD using a three-step approach. *ACM Transactions of Design Automation of Electronic Systems*, 2002. **7**(3): pp. 413–429.

20. H. A. N. Hong-lei, L. I. U. Wen-ju et al. An efficient algorithm of hardware/software partitioning and scheduling on MPSoC. *Computer Engineering and Science*, 2011. **33**(9): pp. 616–8.

21. T. Wiangtong, P. Y. K. Cheung et al. Comparing three heuristic search methods for functional partitioning in hardware-software codesign. *Design Automation for Embedded Systems*, 2002. **6**(4): pp. 425–449.

22. C. Wang, X Li et al. Architecture support for task out-of-order execution in MPSoCs. *IEEE Transactions on Computers*, 2015. **64**(5): pp. 1296–1310.

23. C. Wang, X. Li et al. MP-Tomasulo: A Dependency-aware automatic parallel execution engine for sequential programs. *ACM Transactions of Architecture and Code Optimization*, 2013. **10**(2): pp. 1–24.

24. C. Li, X. Li et al. A dependency aware task partitioning and scheduling algorithm for hardware-software codesign on MPSoCs. In *Proceedings of the 12th International Conference on Algorithms and Architectures for Parallel Processing*, Fukuoka, Japan, 2012: pp. 332–346.

25. R. P. Dick, D. L. Rhodes et al. TGFF: Task graphs for free. In *Proceedings of the 6th International Workshop on Hardware/Software Codesign*, Seattle, WA, 1998: pp. 97–101.

Index

Note: Locator followed by '*f*' and '*t*' denotes figure and table in the text

Printed and bound by CPI Group (UK) Ltd, Croydon, CR0 4YY

29/10/2024

01780548-0002